Collins

Student Support Materials for AQA

A-level Year 2 Chemistry

Paper 2 Organic chemistry and relevant physical chemistry topics

Authors: Colin Chambers, Geoffrey Hallas, Andrew Maczek, David Nicholls, Rob Symonds, Stephen Whittleton

William Collins' dream of knowledge for all began with the publication of his first book in 1819.

A self-educated mill worker, he not only enriched millions of lives, but also founded a flourishing publishing house. Today, staying true to this spirit, Collins books are packed with inspiration, innovation and practical expertise. They place you at the centre of a world of possibility and give you exactly what you need to explore it.

Collins. Freedom to teach

HarperCollins Publishers
The News Building
1 London Bridge Street
London SE1 9GF

HarperCollins Publishers
Macken House, 39/40 Mayor Street Upper,
Dublin 1,
DO1 C9W8,
Ireland

Browse the complete Collins catalogue at www.collins.co.uk

10 9 8 7 6 5 4

ISBN 978-0-00-818951-8

www.collins.co.uk

A catalogue record for this book is available from the British Library

Thanks to John Bentham and Graham Curtis for their contributions to the previous editions.

Commissioned by Gillian Lindsey
Edited by Alexander Rutherford
Project managed by Maheswari PonSaravanan at Jouve
Development by Tim Jackson
Copyedited and proof read by Janette Schubert
Typeset by Jouve India Private Limited
Original design by Newgen Imaging
Cover design by Angela English
Printed and bound in the UK by Ashord Colour Press Ltd
Cover image © Shutterstock/isaravut

Contents

Introduction

Book 4 covers the Physical Chemistry in section 3.1.9 and the Organic Chemistry in sections 3.3.7 to 3.3.16.

Questions in each of the A-level papers will assess not only your knowledge and understanding of particular sections of the specification but also your ability to draw together this knowledge and understanding and apply it in new and unfamiliar contexts.

A-level Paper 2 can include questions on:

- the Physical Chemistry in sections 3.1.2 to 3.1.6
- all the Organic Chemistry listed in section 3.3.

Questions will also assess your practical experience, including knowledge and understanding of the required practicals.

The Physical Chemistry topics in sections 3.1.2 to 3.1.4 and 3.1.6 are covered in Book 1. The Physical Chemistry topic in section 3.1.5 and the Organic Chemistry topics in sections 3.3.1 to 3.3.6 have been covered in Book 2. These topics are assessed in AS Paper 2 but also form part of the end-of-course A-level assessment and may appear in A-level Paper 2.

A-level Paper 3 can include questions on any of the topics in Books 1, 2, 3 and 4. The main emphasis in this final paper will be on synoptic topics, including practical work and data analysis. The objective questions in Paper 3 will aim, together with Papers 1 and 2, to complete a comprehensive coverage of the whole of the A-level specification.

3.1 Physical chemistry

3.1.9 Rate equations

Essential Notes

The SI units of reaction rate are: mol dm^{-3} s^{-1}.

The SI unit of time is the *second*. The general units for rate are *concentration time*$^{-1}$ so, for some very slow reactions, the rate may be expressed as mol dm^{-3} min^{-1} or mol dm^{-3} hr^{-1}.

3.1.9.1 Rate equations

Rate of a chemical reaction

Definition

The rate of a chemical reaction (rate of reaction) is the change in concentration of a substance in unit time.

This definition is also found in *Collins Student Support Materials: AS/A-Level year 1 – Organic and Relevant Physical Chemistry*, section 3.1.5.3.

The rate depends on the temperature of the reaction (see this book, section 3.1.9.2) and also on the concentrations of the reagents involved. However, the actual relationship at a fixed temperature between the rate of reaction and the reactant concentrations cannot be predicted from the overall chemical equation. Take, for example, a reaction for which the overall chemical equation is:

$$A + 2B \rightarrow C$$

Although the rate of this reaction may well depend on either or both of the reactant concentrations [A] and [B], the rate cannot be assumed to be *directly* proportional (mole per mole) to these concentrations. Instead, the rate is given by the expression:

Notes

n and *m* are determined from experimental data.

$$rate \propto [A]^m[B]^n$$

If the rate expression is modified so that the proportional sign is changed into an *equals sign*, it becomes a **rate equation**, to which is added a constant of proportionality, *k*, called the **rate constant** (or velocity constant).

Notes

The value of the rate constant, *k*, varies with temperature.

Definition

The rate equation** expresses the relationship between the rate of reaction and the concentrations of reactants; the constant of proportionality in the rate equation is called the **rate constant.

At a given temperature, *k* is constant so that:

$$rate = k[A]^m[B]^n$$

The powers *m* and *n* are usually integral, most commonly 0, 1 or 2, and are called the **orders of reaction** with respect to the reactants A and B. The *values* of *m* and *n* can never be inferred from the coefficients (numbers of reacting moles) in the stoichiometric equation; the order with respect to a given component is always deduced from experiment.

Definition

*The overall **order of a reaction** is the sum of the powers of the concentration terms in the rate equation.*

In the above case, $(m + n)$ is the sum of the powers of the concentration terms, so the overall reaction has order $(m + n)$.

Zero-order reactions

Consider a reaction for which the rate equation is:

$$rate = k[A]^x$$

If x is zero in this equation, then:

$$rate = k$$

which means that the rate of reaction is *always* constant and independent of the concentration of A, because $[A]^0 = 1$.

If the rate of reaction is constant (Fig 1), a graph of [A] against time is a straight line (Fig 2).

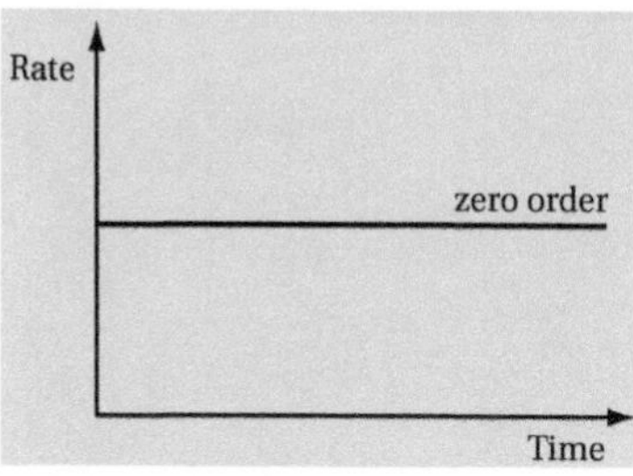

Fig 1
Rate against time for a zero-order reaction

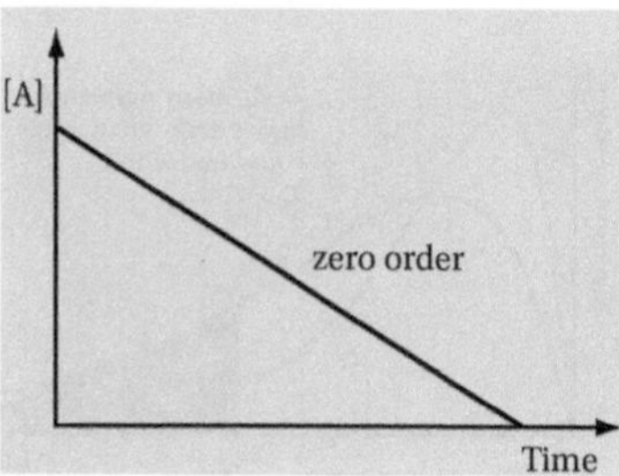

Fig 2
Concentration against time for a zero-order reaction

First-order reactions

If x is 1 in the equation above, the rate equation becomes:

$$rate = k[A]$$

and the reaction is **first order** with respect to A. If [A] doubles, the rate doubles.

Fig 3 shows how concentration varies with time for a first-order reaction. The first-order rate constant, k, has the units s^{-1}, as can be seen by rearranging the rate equation to give:

$$k = \frac{rate}{[A]}$$

in which the units of concentration can be cancelled:

$$\frac{\cancel{(mol\ dm^{-3})}\ s^{-1}}{\cancel{(mol\ dm^{-3})}} = s^{-1}$$

Second-order reactions

The rate equation for a second-order reaction could be:

$$rate = k[A]^2$$

If so, the reaction is **second order** with respect to A.

Alternatively, the rate equation could be:

$$rate = k[A][B]$$

If so, the reaction is *first order* with respect to both A and B and the overall order is $(1 + 1) = 2$.

Fig 3 also illustrates how concentration varies with time for a second-order reaction.

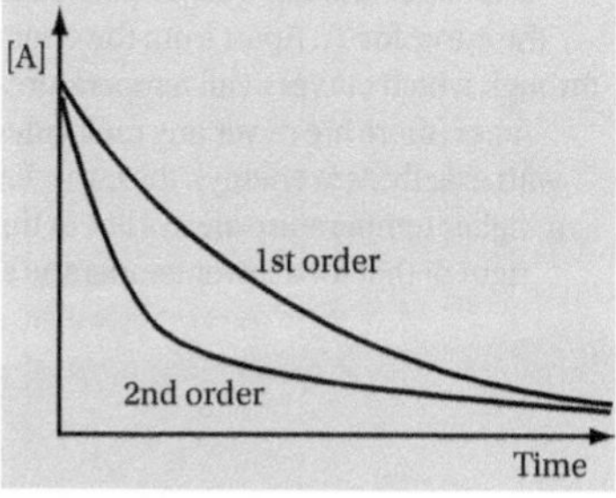

Fig 3
Concentration against time for first- and second-order reactions

A second-order rate constant has units $mol^{-1}\ dm^3\ s^{-1}$. Rearranging the rate equation:

$$k = \frac{rate}{[A][B]}$$

allows units of concentration to cancel:

$$\frac{\cancel{(mol\ dm^{-3})}\ s^{-1}}{\cancel{(mol\ dm^{-3})}(mol\ dm^{-3})} = \frac{s^{-1}}{(mol\ dm^{-3})} = mol^{-1}\ dm^3\ s^{-1}$$

Higher-order reactions

A general form of the rate equation for two reactants A and B is:

$$rate = k[A]^m[B]^n$$

If $n = 1$ and $m = 2$ (or vice-versa) the reaction is **third order** overall.

If $n = 2$ and $m = 2$ (or any other integers that add up to 4) the reaction is **fourth order** overall.

The units for third-order and fourth-order rate constants can be derived using the method shown above. Table 1 on page 12 shows the units of all orders of reaction from order 0 to order 4.

The experimental determination of orders of reaction and rate equations is dealt with in the next section.

The effect of changes in temperature on rate constant

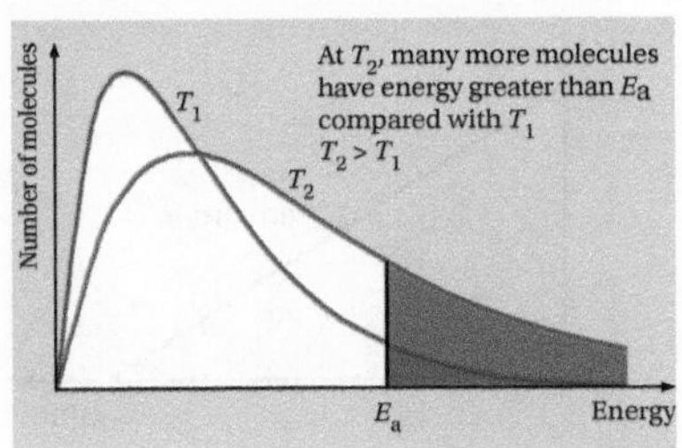

Fig 4
Molecules with energies greater than E_a at different temperatures. The curve for T_2, the higher temperature, is broader and has a lower peak than the curve for T_1. Apart from the origin through which curves at all temperatures pass (there are never any molecules with exactly zero energy), the curve for a higher temperature always lies to the right of that for a lower temperature.

An increase in temperature *increases* the *rate* of a reaction. According to kinetic theory, the mean kinetic energy of the particles is directly proportional to the temperature. At higher temperatures, particles have more energy; they move about more quickly, there are more collisions, and these collisions are more energetic. The *increased energy of the collisions* is a much more important factor in affecting the rate than is the relatively slight *increase in the collision rate* when the temperature is raised.

Fig 4 shows what happens to the distribution of energies in molecules of a gas when the temperature is increased from T_1 to T_2. For a fixed sample of gas, the total number of molecules is unchanged and the total area under the curve remains constant (*Collins Student Support Materials: AS/A-Level year 1 - Organic and Relevant Physical Chemistry,* section 3.1.5.2). To the right of the maximum, the curve at higher temperature T_2 lies above the one at lower temperature T_1; at the higher temperature there are more molecules with greater energy than there are at the lower temperature.

Particles will react only if, on collision, they have more than a minimum amount of energy known as the **activation energy**.

Definition

*The **activation energy** of a reaction is the minimum energy required for the reaction to occur.*

Fig 4 shows that, when the activation energy for a reaction is E_a, the number of molecules with energy in excess of E_a is much greater at the higher temperature T_2 than at the lower temperature T_1. The number of collisions between molecules with sufficient energy to react, i.e. the number of productive collisions, and therefore the rate of reaction, will be greater at the higher temperature. Consequently, quite small temperature increases can lead to very large increases in rate (see Fig 5).

This increase in rate arises because the value of the *rate constant, k,* gets bigger when the temperature increases. The *rate equation*:

$$rate = k[A]^m[B]^n$$

has only one possible temperature-dependent feature – the rate constant; the concentrations of reactants do not change with temperature. Thus it is the rate constant k that is affected by the the number of effective collisions in unit time; the number of such collisions is itself dependent on the temperature of the reaction.

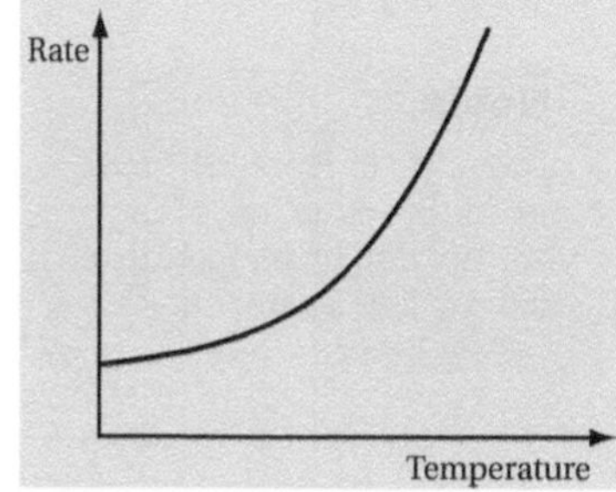

Fig 5
The exponential increase of rate with temperature

The curve in Fig 5, which shows the increase in reaction rate with temperature, also matches the increase in the value of the rate constant as the temperature is increased. The increase in the rate constant with temperature is said to be exponential, which means that every time the temperature increases by a certain number of degrees, the rate constant increases by a fixed factor (two-fold, say, or even ten-fold) depending on the size of the temperature increase.

In many chemical reactions that occur reasonably quickly near room temperature (typically biological reactions), a temperature rise of 10 °C increases the rate of reaction by a factor of about two.

The relationship between rate constant, temperature and activation energy

The overwhelming majority of rates of chemical reaction *increase* as the temperature *increases*. The increase in the value of the reaction rate constant k, as the temperature T is raised, can be expressed using the **Arrhenius equation.**

Notes

The Arrhenius equation, a rearranged version and the value of R will be given when required.

Definition

The Arrhenius equation provides a quantitative relationship between the reaction rate constant and the reaction temperature:

$$k = Ae^{-E_a/RT}$$

A is a constant, called the Arrhenius constant, E_a is the activation energy, T is the temperature in kelvin, R is the gas constant and e is a mathematical constant.

Notes

e is the base of natural **logarithms**, so ln x means the same as $\log_e x$

Taking logarithms to the base e ($\log_e$ or ln) of each side of the Arrhenius equation and rearranging gives the equation in a different form:

$$\ln k = \frac{-E_a}{RT} + \ln A$$

This expression is of the type:

$$y = mx + c$$

which represents a linear graph where the slope is m and the intercept is c.

So, just as a graph of y against x produces a straight-line graph of slope m, a graph of ln k against $1/T$ will produce a straight-line graph of gradient $\frac{-E_a}{R}$.

Thus values of E_a and A can be determined.

Notes

If logarithms to the base 10 are used, after taking logarithms and rearranging, the expression then becomes:

$$\log_{10} k = \frac{-E_a}{2.303RT} + \log_{10} A$$

and a graph of $\log_{10} k$ against $1/T$ will produce a straight line graph of gradient $\frac{-E_a}{2.303R}$.

Notes

The units of R are J K^{-1} mol^{-1} so, in this equation, after multiplying by temperature the units of E_a are J mol^{-1}.

Notes

As shown in Fig 7, the gradient (and, by inference, also the rate) is negative. To remove this anomaly, the rate of reaction is measured as:

rate = + gradient for product increase
= – gradient for reactant decrease

This procedure ensures that the quoted reaction rate is always positive.

Notes

Measuring the rate of reaction by an initial rate method and by a continuous monitoring process are required practical activities.

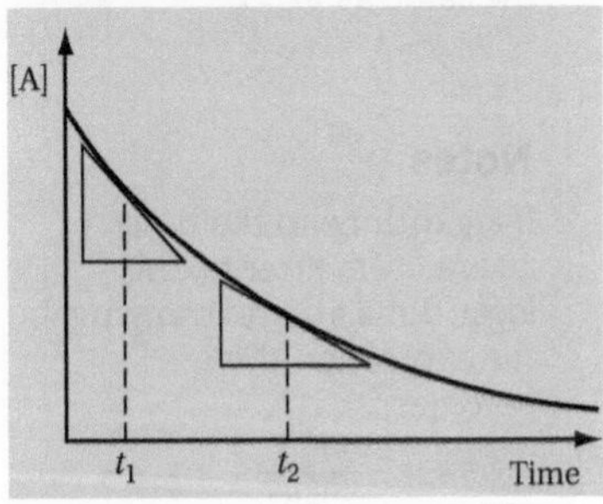

Fig 7
The gradient of the concentration-time curve measures the rate of reaction. The rate is higher at the earlier time (t_1), and falls as the reaction proceeds, eventually falling to zero

Example

A reaction was studied at a number of temperatures and the following results were obtained.

Temperature / K	400	444	500	571	667	800
Rate constant / mol^{-1} dm^3 s^{-1}	0.0714	0.1075	0.1612	0.2466	0.3753	0.562

Since $k = Ae^{-E_a/RT}$ and $\ln k = -E_a/RT + \ln A$, a plot of ln k against $1/T$ gives a straight line of gradient $-E_a/R$.

The values plotted in Fig 6 were:

Temperature / K	400	444	500	571	667	800
Rate constant / mol^{-1} dm^3 s^{-1}	0.0714	0.1075	0.1612	0.2466	0.3753	0.562
1/T	0.00250	0.00225	0.00200	0.00175	0.00150	0.00125
ln k	–2.639	–2.230	–1.825	–1.400	–0.980	–0.576

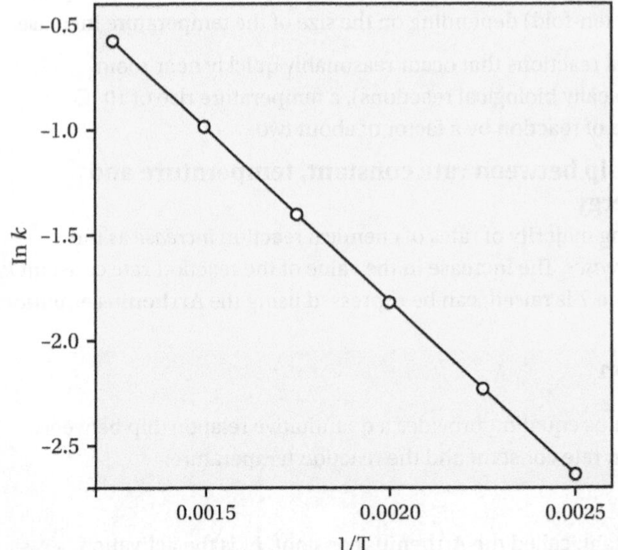

Fig 6 A plot of ln k versus $1/T$.

From the graph in Fig 6, gradient $= \dfrac{(-2.640)-(-0.575)}{0.00250-0.00125} = \dfrac{-2.064}{0.00125} = -1.65 \times 10^3$ K

Since gradient $= -E_a/R$, $E_a = -(\text{gradient} \times R) = -(-1.65 \times 10^3 \times 8.31)$
$= 13720$ J mol^{-1}

So, activation energy, $E_a = 13.7$ kJ mol^{-1}

3.1.9.2 Determination of rate equation

As a reaction proceeds, the *rate* of reaction at fixed temperature decreases because the *concentrations* of the reagents fall as they are being used up. The value of the rate at a particular time can be found by measuring the gradient at that time on a graph of concentration against time (see Fig 7). The rate at the start of the reaction, when the initial concentrations of the reagents are known exactly, is called the **initial rate**.

When the experiment is repeated using different initial concentrations of reagents, the initial rate changes. By recording how different initial concentrations affect the initial rate, chemists can derive the rate equation for a reaction.

Example

The following data were obtained for the reaction:

$$2P + Q \rightarrow R + S$$

Experiment	**Initial [P]/mol dm^{-3}**	**Initial [Q]/mol dm^{-3}**	**Initial rate/mol dm^{-3} s^{-1}**
1	0.5	0.5	0.002
2	1.0	0.5	0.008
3	1.0	1.0	0.008
4	1.5	1.5	0.018

Determine the order of reaction with respect to components P and Q and deduce the units of the rate constant.

Method

Consider pairs of experiments in which the concentrations of one of the reagents remains constant. This approach establishes the order with respect to the other reagent.

Consider experiments 1 and 2:

[Q] remains constant and [P] is doubled.

The rate increases by a factor of (0.008/0.002) = 4 or 2^2 so:

$rate \propto [P]^2$

and the reaction is **second order** with respect to P.

Consider experiments 2 and 3:

[Q] is doubled and [P] is kept constant. The rate is unchanged, i.e. the rate is independent of the concentration of Q so:

$rate \propto [Q]^0$

and the reaction is **zero order** with respect to Q.

Overall:

$rate \propto [Q]^0[P]^2$

so the overall order is 0 + 2 = 2

Hence the rate equation is:

$rate = k[P]^2$

Comment

The reaction is **second order** overall. Derive the units of this rate constant without referring back to the previous page. The answer is in Table 1 below.

Notes

The data given for Experiment 4 are not needed to derive the answer, but can be used as a consistency check. In each of the experiments, the value of k (which is found from the expression $rate/[P]^2$) should be the same. Consistency for Experiments 1 and 4 requires that $0.002/(0.5)^2$ and $0.018/(1.5)^2$ should be the same. Check to see if they are.

There is a very simple pattern to the units of the rate constant and the order of the reaction. This is shown in Table 1 below. To understand how the units of the rate constant can be derived, it is enough to recall that a reaction of order $n + m$ has the rate equation:

$rate = k \times (concentration)^{n+m}$

Table 1
Order of reaction and the units of the rate constant

Order of reaction ***= n + m***	***Rate equation*** ***rate = k × [reactants]$^{n + m}$***	***Units of the*** ***rate constant***
0	rate = k × concentration^0	mol dm^{-3} s^{-1}
1	rate = k × concentration^1	s^{-1}
2	rate = k × concentration^2	mol^{-1} dm^3 s^{-1}
3	rate = k × concentration^3	mol^{-2} dm^6 s^{-1}
4	rate = k × concentration^4	mol^{-3} dm^9 s^{-1}

It is by considering kinetic data that the orders of reactions can be deduced. Information of this kind is very important in deciding how to maximise the rate of a reaction, for example in an industrial process. Increasing the concentration of Q in the example above, for instance, would have no effect on the rate and to do so would be a waste of money.

Notes

The *initiation* step involves only a single molecule reacting on its own; this is called a *unimolecular step.*

Both the *propagation steps* involve two species reacting together; these are called *bimolecular steps.*

Termolecular steps (three species reacting simultaneously) are possible, but extremely rare.

Reaction mechanisms and the rate-determining step

Reaction mechanisms

Chemical reactions rarely occur by the simple and straightforward route suggested by the overall stoichiometric equation. Most reactions occur in two or more steps which, when combined, produce the equation for the overall reaction. One of the major tasks of reaction kinetics lies in providing evidence to support or refute the validity of such proposed reaction steps.

A proposed sequence of simple reaction steps is known as a **reaction mechanism.**

Definition

***The reaction mechanism** for a reaction consists of a proposed sequence of discrete chemical reaction steps that can be deduced from the experimentally observed rate equation.*

Notes

Although the rate equation for a reaction cannot be deduced from the stoichiometry of the reaction equation (the rate equation is always derived experimentally), the rate of any step in a mechanism is always proportional to the concentrations of the reactant species in that step, raised to the appropriate powers.

Thus, in the chlorination chain reaction, it is possible to write for the initiation step:

$rate = k_1[Cl_2]$

which is unimolecular, and therefore a first-order step, and for the first propagation step:

$rate = k_2[CH_4][Cl\bullet]$

which is bimolecular, and therefore a second-order step.

Reactions that occur in steps, and reaction mechanisms, have already been introduced in *Collins Student Support Materials: AS/A-Level year 1 – Organic and Relevant Physical Chemistry*, section 3.3.2.4, where a **free-radical substitution mechanism** is invoked to account for the observed chain-reaction kinetics of the direct chlorination of methane:

Initiation: $Cl_2 \rightarrow 2Cl\bullet$

Propagation: $Cl\bullet + CH_4 \rightarrow \bullet CH_3 + HCl$

$\bullet CH_3 + Cl_2 \rightarrow CH_3Cl + Cl\bullet$

Overall: $CH_4 + Cl_2 \rightarrow CH_3Cl + HCl$

The rate-determining step

In some reactions, the experimental rate equation seems directly related to the process as written in the overall stoichiometric equation. A good example of this is the hydrolysis of 1-bromobutane:

$CH_3CH_2CH_2CH_2Br + OH^- \rightarrow CH_3CH_2CH_2CH_2OH + Br^-$

for which the experimental rate equation is:

$$rate = k[CH_3CH_2CH_2CH_2Br][OH^-]$$

The second-order rate equation suggests a single-step nucleophilic substitution mechanism involving direct collision of a hydroxide ion with a molecule of bromobutane; no further explanation is needed.

However, in the superficially identical hydrolysis of 2-bromo-2-methylpropane:

$$(CH_3)_3CBr + OH^- \rightarrow (CH_3)_3COH + Br^-$$

the experimental rate equation is:

$$rate = k[(CH_3)_3CBr]$$

and the reaction is *first order*, clearly requiring a different *reaction mechanism*.

A proposed reaction mechanism that would fit this rate equation is:

$$(CH_3)_3CBr \xrightarrow{\text{slow}} (CH_3)_3C^+ + Br^-$$

$$(CH_3)_3C^+ + OH^- \xrightarrow{\text{fast}} (CH_3)_3COH$$

Definition

***The rate-determining step** is the slowest step in a multi-step reaction sequence; it dictates the overall rate of reaction.*

Essential Notes

In a process that involves a sequence of steps, each dependent on the preceding one, the overall rate of conversion is limited by the speed of the slowest step.

The first, slow step involving the formation of a carbocation determines the overall rate of reaction. It is called the **rate-determining step**.

The rate-determining step above involves only a single parent molecule; this step can be written as:

$$(CH_3)_3CBr \xrightarrow{\text{slow}} (CH_3)_3C^+ + Br^- \qquad rate = k_1[(CH_3)_3CBr]$$

and the second step can be written as:

$$(CH_3)_3C^+ + OH^- \xrightarrow{\text{fast}} (CH_3)_3COH \qquad rate = k_2[(CH_3)_3C^+][OH^-]$$

However, the kinetics of this step are of no interest, as it can proceed only as fast as the carbocation is formed and then speedily consumed.

The overall reaction is:

$$(CH_3)_3CBr + OH^- \rightarrow (CH_3)_3COH + Br^- \qquad rate = k_1[(CH_3)_3CBr]$$

According to this proposed scheme, the reaction is first order (as found experimentally), so the proposed reaction mechanism is in accord with the observed kinetics.

This is an example of a reaction mechanism where the rate-determining step is the first step in the sequence, so only the reactants in this step can appear in the rate equation. In other cases (mentioned below) the rate-determining step follows after other fast steps; in such cases, the species involved in these fast steps may well appear in the experimental rate equation.

Notes

It is likely that both mechanisms (the second-order as well as the first-order) are in play in the hydrolysis of both compounds.

For 1-bromobutane, the dissociation into ions will have much higher activation energy than will the approach of a hydroxide ion to the C—Br carbon, which is relatively exposed. So the second-order mechanism will dominate.

However, for 2-bromo-2-methylpropane, the interference of the three quite bulky methyl groups will play a significant role in raising the bimolecular activation energy, as also will their influence in stabilising the tertiary carbocation.

Notes

An inescapable consequence of the rate-determining step is that the only species that can figure in the final rate equation are those that appear as reactants in steps up to and including the rate-determining step.

Example

It has been suggested that a reaction of nitrogen dioxide with carbon monoxide can occur in the exhaust gases of motorcars. Kinetic experiments show that the overall reaction is second order in NO_2 and zero order in CO. Suggest a mechanism to account for the observed kinetics and indicate what steps might be taken to support this mechanism.

Method

First write an equation for the overall reaction:

$NO_2(g) + CO(g) \rightarrow CO_2(g) + NO(g)$ $\quad$ $rate = k[NO_2]^2$

Next look for a *rate-determining step* that involves two molecules of NO_2

A possible candidate is:

$2NO_2 \xrightarrow{slow} N_2O_4$ $\quad$ known reaction, known product

$N_2O_4 + CO \xrightarrow{fast} CO_2 + NO + NO_2$ $\quad$ required products, NO_2 recycled

$NO_2(g) + CO(g) \rightarrow CO_2(g) + NO(g)$ $\quad$ overall, as required

Further investigations

If this is the mechanism, it should be possible to detect dinitrogen tetroxide during the reaction. This could be done spectroscopically. But no N_2O_4 is found at the temperature of the experiment; instead, spectroscopic investigations detect the presence of a short-lived intermediary, NO_3.

Comment

The mechanism above is very unlikely, as no N_2O_4 is found. Also, it takes no account of the presence of NO_3. So, following the scheme above, it is necessary to look for another *rate-determining step* that involves two molecules of NO_2 and also one that produces NO_3.

$2NO_2 \xrightarrow{slow} NO_3 + NO$ $\quad$ detected product formed

$NO_3 + CO \xrightarrow{fast} CO_2(g) + NO_2$ $\quad$ required products, NO_2 recycled

$NO_2(g) + CO(g) \rightarrow CO_2(g) + NO(g)$ $\quad$ overall, as required

It cannot be proved that this is the actual mechanism, but it corresponds to all the known facts whereas the earlier one does not.

Notes

This example has been given in order to illustrate how it is that chemists approach problems, and how they try to formulate hypotheses to solve them.

Nothing as demanding as this will be expected in the AQA A2 examination.

Notes

This reaction is unusual in that, unlike other chemical reactions, the rate decreases as temperature is increased. This interesting observation lies outside the scope of the specification.

The example below illustrates how kinetic and other data can be used to make proposals about reaction mechanisms and the rate-determining step.

In cases where the rate-determining step is preceded by other steps in the mechanism, the situation becomes more complicated. An example of this is the oxidation of nitrogen(II) oxide, which exhibits third-order kinetics:

$2NO(g) + O_2(g) \rightarrow 2NO_2(g)$ $\quad$ $rate = k[NO]^2[O_2]$

Notes

Knowledge of the oxides of nitrogen lies outside the scope of the AQA A-level specification.

While it is possible that this reaction occurs by a direct mechanism involving the simultaneous collision of three molecules, this is extremely unlikely because ternary (three-body) collisions are extremely rare. So a multi-step mechanism needs to be sought.

A possible mechanism that accounts for the known facts involves the formation of a dimer, N_2O_2, as is shown below:

$$2NO \xrightarrow{\text{fast}} N_2O_2 \qquad rate = k_{1f}[NO]^2$$

$$N_2O_2 \xrightarrow{\text{fast}} 2NO \qquad rate = k_{1b}[N_2O_2]$$

$$N_2O_2 + O_2 \xrightarrow{\text{slow}} 2NO_2 \qquad rate = k_2[N_2O_2][O_2]$$

The fact that the overall reaction is first order in oxygen makes it inevitable that the third equation must be the rate-determining step, as it alone involves the concentration of oxygen. Details of this mechanism and how it can lead to an (experimental) rate equation lie well beyond the scope of the specification. However, an outline is given in the Notes, shown opposite.

Notes

The first two reactions are in equilibrium, with forward (f) and backward (b) rate constants. Forward and backward rates are equal, so:

$k_{1f}[NO]^2 = k_{1b}[N_2O_2]$

and $[N_2O_2] = \frac{k_{1f}}{k_{1b}}[NO]^2$

Substitution yields:

$$rate = \frac{k_{1f}}{k_{1b}} k_2[NO]^2[O_2]$$

as required by experiment.

3.3 Organic chemistry

3.3.7 Optical isomerism

Structural isomerism was considered in (see *Collins Student Support Materials: A-Level year 1 - Organic and Relevant Physical Chemistry*, section 3.3.1.3). and includes chain isomerism, position isomerism and functional group isomerism:

- **Chain isomerism** occurs when there are two or more ways of arranging the carbon skeleton of a molecule.
- **Position isomerism** occurs when the **isomers** have the same carbon skeleton, but the functional group is attached at different places on the chain.
- **Functional group isomerism** occurs when different **functional groups** are present in compounds which have the same molecular formula.

Stereoisomerism

The two types of **stereoisomerism** are ***E–Z* stereoisomerism** (see *Collins Student Support Materials: A-Level year 1 - Organic and Relevant Physical Chemistry*, section 3.3.1.3) and **optical isomerism**.

Definition

***Stereoisomers** are compounds which have the same structural formula but the bonds are arranged differently in space.*

Notes

When three or four different groups are attached to a C=C bond, Cahn–Ingold–Prelog (CIP) priority rules are used. These rules assign the priorities for use when naming such compounds. They involve looking at the two atoms attached directly to the left-hand carbon of a double bond, and giving priority to the atom with the highest atomic number. Similarly, for the right-hand carbon of the double bond, the atom with the highest atomic number is assigned the highest priority. CIP priority rules are discussed in more detail in *Collins Student Support Materials: A-Level year 1 – Organic and Relevant Physical Chemistry*, section 3.3.1.3.

E–Z stereoisomerism

Because of restricted rotation at the C=C bond, *Z* or *cis* and *E* or *trans* forms occur when there is suitable substitution:

Essential Notes

E–Z stereoisomerism is also known as geometrical or *cis–trans* isomerism.

It is not possible to have *E–Z* stereoisomerism when there are two identical groups joined to the same carbon atom in a double bond. Restricted rotation about a C=C bond arises due to the interaction between the two adjacent p-orbitals of the carbon atoms, forming a π-bond. Disruption of the π-bond requires significantly more energy than is available at room temperature, so that rotation does not occur readily.

Optical isomerism

When four different atoms or groups are attached to a carbon atom, the molecule has no centre of symmetry, plane of symmetry or axis of symmetry. The molecule is said to be **chiral** and to possess an **asymmetric carbon atom.** Two tetrahedral arrangements in space are possible so that one is the mirror image of the other; 2-hydroxypropanenitrile (see this book, section 3.3.8) has two such stereoisomers (see Fig 8).

Stereoisomers of this kind are known as **enantiomers.** It is not possible to superimpose one enantiomer on the other. Enantiomers have the same physical properties except for their effect on the plane of plane-polarised light and, because of this difference, are said to be **optically active.**

Essential Notes

Note that the dashed bond is behind the plane of the page, whereas the wedge bond is in front of this plane.

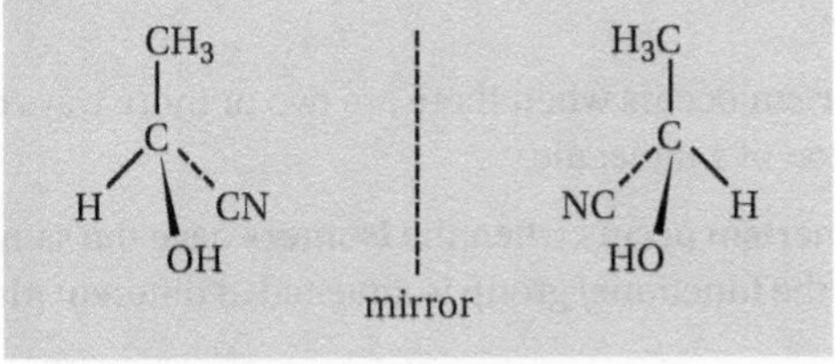

Fig 8
Stereoisomers of 2-hydroxypropanenitrile

When plane-polarised light, which is made up of waves vibrating in one plane only, passes through a solution of a chiral compound, the light emerges with its direction of polarisation changed. One enantiomer will rotate the plane of plane-polarised light in a clockwise direction; it is termed (+) or dextrorotatory. Its mirror-image form will rotate the plane of plane-polarised light by the same amount in a counterclockwise direction; it is termed (–) or laevorotatory. A mixture of equal amounts of both enantiomers is optically inactive, because the two effects cancel out. Such a mixture is called a **racemic mixture** or a **racemate.**

Many naturally-occurring molecules exist as single enantiomers, notably most amino acids (see this book, section 3.3.13.1), such as 2-aminopropanoic acid, *alanine*, $CH_3CH(NH_2)COOH$. The chemical properties of enantiomers are identical except in reactions with other optically active substances. Because enzymes are stereospecific, they can distinguish between enantiomers and catalyse the reactions of only one of a pair of isomers.

2-Hydroxypropanenitrile can be prepared (Fig 11, page 19, R = CH_3) by the reaction of potassium cyanide with KCN followed by dilute acid, with ethanal (see this book, section 3.3.8). Because the carbonyl group is planar, attack by the **nucleophile** CN^- is equally likely from either side of the plane, leading to the formation of a racemate (Fig 9). Hydrolysis of this product causes the CN group to be converted into COOH. The final product is the racemate of 2-hydroxypropanoic acid, commonly called *lactic acid*. This (±)-mixture is found in sour milk; the naturally occurring (+)-enantiomer is formed during the contraction of muscles.

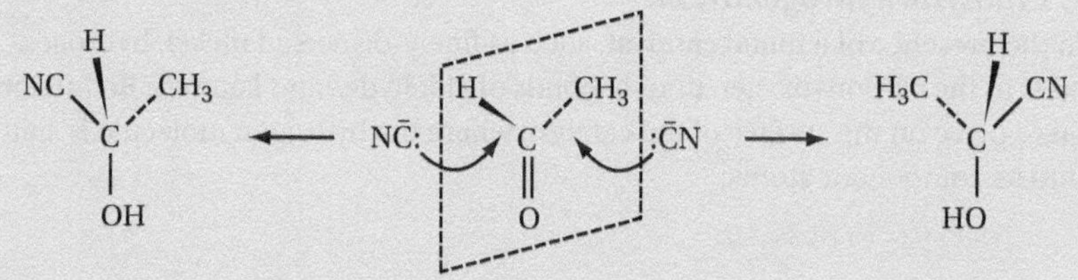

Fig 9
Racemate formation

Chiral drugs

Approximately half of the commercially-available drugs contain at least one chiral centre. Although drugs extracted from natural sources, e.g. quinine, are single enantiomers, most synthetic products are obtained as racemic mixtures. Because of the inherent difficulty in, and high cost of, separating the enantiomers, with few exceptions, synthetic drugs have been marketed as racemates. However, enantiomers can have unequal degrees of the same physiological activity or very different activities.

The drug thalidomide was used originally, as the racemate, to combat morning sickness in pregnant women. One enantiomer proved to be a potent teratogen (causes malformation of a fetus), leading to severe birth defects. The other enantiomer is not teratogenic but, unfortunately, in this particular case, the two stereoisomers can interconvert **in vivo** (within the human body).

The separation of enantiomers, known as **resolution**, can be achieved by reaction of the mixture with a chiral reagent, resulting in products with different physical properties, e.g. solubility. Enantiomers can also be separated directly by chromatography (see this book, section 3.3.16).

Notes

Receptors on the surface of nerve cells in the nose are sometimes able to distinguish between certain pairs of enantiomers. For example, the smells of the enantiomers of carvone, a naturally-occurring terpene, are very different. (–)-carvone is found in spearmint oil, whereas (+)-carvone is the main constituent of caraway seed oil. Each enantiomer fits a different receptor in the nose.

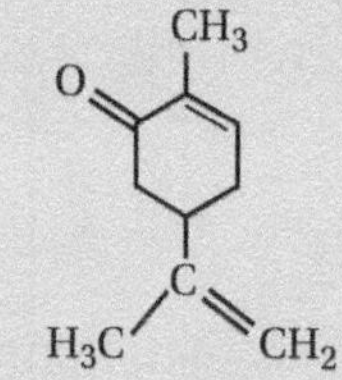

carvone

Notes

Drug companies strive to develop single-enantiomer products, which often possess enhanced properties. The popular analgesic ibuprofen acts much faster as the (+)-enantiomer than when administered as the racemate.

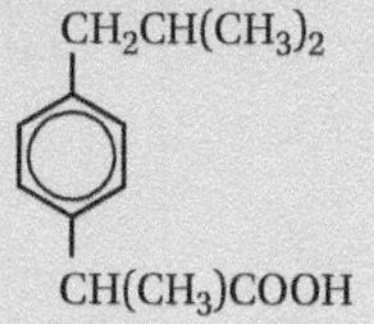

ibuprofen

Exercise: Can you identify the chiral carbon in ibuprofen? [Look for a saturated carbon atom with four different groups attached.]

Essential Notes

The maximum number of stereoisomers for a compound with n chiral centres is 2^n; this number is reduced by symmetrical substitution.

3.3.8 Aldehydes and ketones

Oxidation of aldehydes

The oxidation of aldehydes to carboxylic acids was covered in *Collins Student Support Materials: AS/A-Level year 1 – Organic and Relevant Physical Chemistry*, section 3.3.5.2.

$RCHO + [O] \rightarrow RCOOH$

That section described the use of Fehling's solution or Tollens' reagent to distinguish between aldehydes and ketones (see Table 14 in *Collins Student Support Materials: AS/A-Level year 1 – Organic and Relevant Physical Chemistry*, section 3.3.6.1).

Notes

Note that reduction of an unsymmetrical ketone gives rise to a racemic product, e.g. $CH_3COCH_2CH_3$ is converted into $(\pm)CH_3CH(OH)CH_2CH_3$.

Reduction of aldehydes and ketones

Reduction of the carbonyl group in aldehydes and ketones leads to the formation of primary alcohols and secondary alcohols, respectively.

1. Catalytic hydrogenation

In the presence of a metal **catalyst**, such as finely-dispersed nickel, hydrogen adds to the carbon–oxygen double bonds of aldehydes and ketones. Reduction takes place on the surface of the catalyst, where the hydrogen molecule is split into its component atoms:

$RCHO + H_2 \rightarrow RCH_2OH$

$RCOR + H_2 \rightarrow RCH(OH)R$

Notes

Note that $NaBH_4$ has no effect on alkenes, neither does $LiAlH_4$, so that selective reduction of the carbonyl group can be achieved in compounds containing both carbon–carbon and carbon–oxygen double bonds. A reducing agent such as hydrogen in the presence of a nickel or a platinum catalyst would reduce all unsaturated linkages.

2. Addition of hydride ion

Carbon–oxygen double bonds are polar since oxygen, being more electronegative than carbon, has a greater share of the bonding electrons between the two atoms. Such groups are readily reduced by reagents which are sources of nucleophilic hydride ions ($:H^-$). Sodium tetrahydridoborate(III) ($NaBH_4$) can be used in aqueous ethanol, but anhydrous conditions are required for the more powerful reducing agent lithium tetrahydridoaluminate(III) ($LiAlH_4$).

The mechanism of hydride-ion reduction involves nucleophilic attack on the electron-deficient ($\delta+$) carbon atom of the carbonyl group, to form an oxyanion which is subsequently protonated by water or a weak acid (Fig 10). This is a nucleophilic addition reaction, with the hydride ion acting as the nucleophile (electron-pair donor).

Fig 10
Hydride reduction of an aldehyde

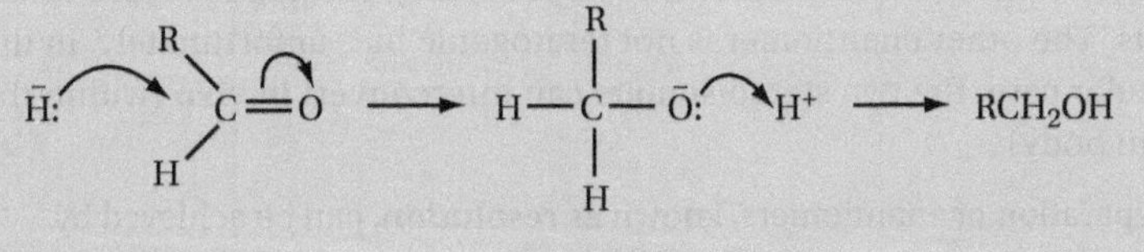

$\bar{H}: + RHC{=}O \longrightarrow H{-}C(R)(H){-}\bar{O}: \;\; H^+ \longrightarrow RCH_2OH$

$RCHO + 2[H] \rightarrow RCH_2OH$

3. Addition of hydrogen cyanide

Hydrogen cyanide adds nucleophilically to aldehydes and ketones to form hydroxynitriles, leading to an increase in the number of carbon atoms. Thus, ethanal is converted into 2-hydroxypropanenitrile:

$CH_3CHO + HCN \rightarrow CH_3CH(OH)CN$

Because hydrogen cyanide is a highly toxic gas, the best way of carrying out the reaction is to use potassium cyanide followed by dilute acid.

In the mechanism, the nucleophilic cyanide ion attacks the carbonyl group to form an oxyanion which then accepts a proton (Fig 11). This is another example of a nucleophilic addition reaction.

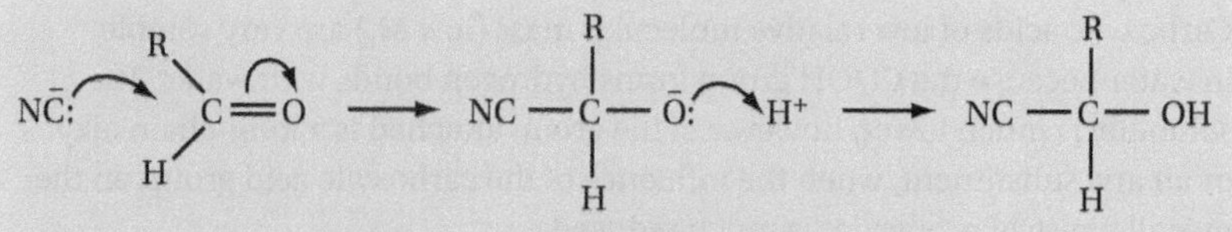

$$N\bar{C}: + R(H)C{=}O \longrightarrow NC{-}C(R)(H){-}\bar{O}: \; H^+ \longrightarrow NC{-}C(R)(H){-}OH$$

Fig 11
Nucleophilic addition of hydrogen cyanide

The reaction product, a hydroxynitrile, is formed as an optically-inactive racemate, because attack by the cyanide nucleophile occurs with equal probability from either side of the planar carbonyl group (see Fig 9). All aldehydes and unsymmetrical ketones will form such mixtures of enantiomers. Symmetrical ketones do not, as their products do not possess an asymmetric carbon atom.

The addition of hydrogen cyanide to aldehydes and ketones is useful in synthesis. Catalytic hydrogenation reduces the CN group to the CH_2NH_2 group (see this book, section 3.3.11.1). 2-Hydroxypropanenitrile is used to illustrate this type of reaction in Fig 12.

Essential Notes

The CN group is converted into the COOH group by acid-catalysed hydrolysis.

$$CH_3CH(OH)CN \xrightarrow[H_2]{Ni} CH_3CH(OH)CH_2NH_2$$

1-aminopropan-2-ol

Fig 12
Reduction of 2-hydroxypropanenitrile

The high toxicity of cyanides requires extreme caution in their use for organic synthesis.

Notes

The use of NaCN or KCN avoids the danger of rapid death caused by inhalation of HCN. Great care is necessary when using these salts. They are extremely toxic reagents and even small amounts, if ingested, can prove fatal.

3.3.9 Carboxylic acids and derivatives

3.3.9.1 Carboxylic acids and esters

Carboxylic acids

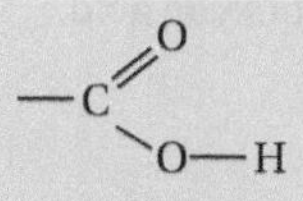

$$-C(=O){-}O{-}H$$

a carboxyl group

Carboxylic acids contain the functional group COOH (see Table 2).

Although this group is made up of a **carb**onyl group (as in aldehydes and ketones) and a hydr**oxyl** group (as in alcohols), the two groups interact so that the properties of carboxylic acids are, in the main, different from those of carbonyl compounds and alcohols.

Essential Notes

$$K_a = \frac{[H^+(aq)][A^-(aq)]}{[HA(aq)]}$$

See *A-Level year 2 – Inorganic and Relevant Physical Chemistry*, section 3.1.12.4.

Table 2
Some simple carboxylic acids

Formula	*Name*
HCOOH	methanoic acid
CH_3COOH	ethanoic acid
CH_3CH_2COOH	propanoic acid
C_6H_5COOH	benzenecarboxylic acid

Notes

The charge on the carboxylate anion is shared equally between the two oxygens, so that the charge on each oxygen is effectively halved; the stabilised anion is less likely to regain a proton than expected:

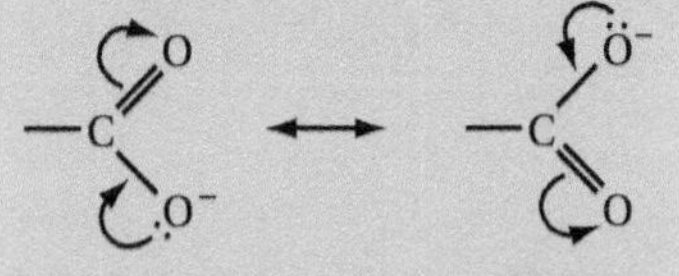

or

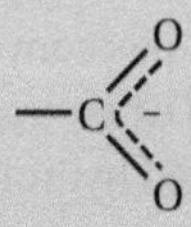

on average

The double-headed resonance arrow should not be confused with the symbol used to denote equilibrium between different species.

Notes

$NaHCO_3$ gives carbon dioxide with a weak acid when in solution and in the solid form.

Na_2CO_3 will give carbon dioxide only when it is used in the solid form as a powder.

Acidity

Carboxylic acids of low relative molecular mass (low M_r) are very soluble in water because the COOH group forms hydrogen bonds with water. The solubility is much lower, however, if the group attached is a long-chain alkyl or an aryl substituent, when the influence of the carboxylic acid group on the overall physical properties is much reduced.

Although the solubility may be high, the acids are only slightly dissociated in water, i.e. they are weak acids. The carbonyl group attracts electrons away from the alcohol group so that the O—H bond is weakened and can break more easily to release a proton and produce a stable carboxylate anion, as in the case of ethanoic acid:

$$CH_3COOH(aq) \rightleftharpoons CH_3COO^-(aq) + H^+(aq)$$

As distinct from strong acids, the amount of this dissociation is quite small. The dissociation constant, K_a, for ethanoic acid is 1.76×10^{-5} mol dm^{-3}, which means that a solution containing 0.1 mol dm^{-3} of the acid is only about 0.3% ionised. By comparison, the O—H bond in an alcohol does not break easily, so that alcohols do not show typical acidic properties.

Carboxylic acids react as normal acids with metals, alkalis and carbonates to form salts, although the reactions are less vigorous than with strong acids.

When carboxylic acids react with sodium hydrogencarbonate, carbon dioxide is evolved. This reaction can be used as a test for carboxylic acids:

$$CH_3COOH + NaHCO_3 \rightarrow CH_3COONa + H_2O + CO_2$$

The salts formed are ionic and are therefore water-soluble.

Esters

Esters contain the functional group COOR, where R is usually an alkyl group. Esters can be formed by the reaction of carboxylic acids with alcohols in the presence of strong-acid catalysts, as in the case of methyl ethanoate (see also acylation):

$$CH_3COOH + CH_3OH \rightleftharpoons CH_3COOCH_3 + H_2O$$

When ethanoic acid and methanol are mixed in the presence of a small quantity of concentrated sulfuric acid, methyl ethanoate is formed as a volatile, sweet-smelling product.

Naming esters

The ionic carboxylate group in a salt of a carboxylic acid is named by substituting *-oate* for the *-oic* ending from the name of the carboxylic acid.

The acid part of an ester is similarly named:

CH_3COO^- is called ethanoate
$CH_3CH_2COO^-$ is called propanoate

The alcohol part of the ester name is written first, as in Table 3.

Name	Formula
methyl propanoate	$CH_3CH_2COOCH_3$
ethyl methanoate	$HCOOCH_2CH_3$
ethyl benzenecarboxylate	$C_6H_5COOCH_2CH_3$

Table 3
Examples of complete names

Uses of esters

1. Solvents

Esters have no free OH groups, so cannot form hydrogen bonds in the same way as carboxylic acids or alcohols; they are therefore much more volatile than carboxylic acids and are almost insoluble in water. Esters are polar, however, and are able to act as solvents for many polar organic compounds; their relatively low boiling points allow them to be separated easily from the less volatile solutes. Typical examples include:

- ethyl ethanoate (b.p. 77 °C), which is commonly used as the solvent in glue, such as polystyrene cement, and in printing inks;
- butyl ethanoate (b.p. 126 °C), which is widely used as a solvent in the pharmaceutical industry, for nitrocellulose, and in many lacquers.

2. Plasticisers

Although the forces between the chains in thermoplastic polymers are weak, the material is often not soft or flexible because the polymer chains cannot move easily over each other. Incorporating plasticisers into the polymer allows the chains some movement and, by adding different amounts of plasticiser, the flexibility of the material can be adjusted. For example, the esters of benzene-1,2-dicarboxylic (*phthalic*) acid or of hexanedioic (*adipic*) acid can constitute up to 50% of some plastics such as PVC (see *Collins Student Support Materials: AS/A-Level year 1 - Organic and Relevant Physical Chemistry*, section 3.3.4.3). Over time these additives escape from the plastic, which then becomes brittle and stiff with age.

3. Perfumes

The strong and pleasant smells of many esters, whether naturally occurring in plants, or synthesised in the laboratory, have made them a major component of perfumes and fragrances.

The following properties of esters make them suitable for inclusion in perfumes:

- they evaporate easily, so they can reach the nose
- they evaporate at different rates depending on their relative molecular mass, so the effect is prolonged
- they do not react with water in sweat to form unpleasant-smelling or skin-irritating compounds
- they are not water soluble and so are not easily washed off
- they are non-toxic and although absorbed by the skin, do not cause irritation or poisoning.

When designing a perfume to give a particular fragrance, the chemist will include 'top notes', the lightest (lowest relative molecular mass) and therefore the first fragrant molecules to evaporate and reach the nose (important in making an initial impression), and 'low notes', the heaviest (highest relative molecular mass) and slowest molecules to evaporate (needed for prolonged effect).

Essential Notes

Ester formation involves the elimination of water. Isotopic labelling with ^{18}O shows that the OH group is lost from the acid and not the alcohol.

When making esters from carboxylic acids, carefully pour the equilibrium mixture formed into an excess of warm water. The remaining carboxylic acid and alcohol dissolve, but the ester is immiscible and floats on the water. The aroma of the ester is no longer contaminated with that of the acid and is easily detected.

Notes

It should be recognised that esters and carboxylic acids are good examples of functional group isomerism (see *Collins Student Support Materials: AS/A-Level year 1 – Organic and Relevant Physical Chemistry*, section 3.3.1.3

Notes

The addition of suitable plasticisers to PVC enables the resulting material to be used as a synthetic leather.

4. Food flavourings

Many esters have sweet, often fruity, smells and are used as artificial fruit flavourings. Some of the common esters used in flavourings are shown in Table 4. Natural fruit flavours are extremely complex mixtures of many esters and carboxylic acids. Artificial flavours using only some of these compounds will inevitably only approximate to the real thing.

Table 4
Esters used as food flavourings

Name	*Flavour*
pentyl ethanoate	pear
2,2-dimethylpropyl ethanoate	banana
octyl ethanoate	orange
ethyl butanoate	rum
pentyl pentanoate	apple

Essential Notes

Solid esters occur in animal fats; vegetable oils are liquid esters.

Hydrolysis of esters

When an ester is heated with alkali, it is hydrolysed to an alcohol and a carboxylate salt, e.g. ethyl ethanoate is hydrolysed by aqueous sodium hydroxide to produce ethanol and sodium ethanoate:

$$CH_3COOC_2H_5 + NaOH \rightarrow C_2H_5OH + CH_3COONa$$

This hydrolysis reaction is used widely with naturally-occurring esters, such as oils and fats, to produce useful products including soaps and glycerol.

Most oils and fats are esters of propane-1,2,3-triol (*glycerol*) with three long-chain carboxylic acids. These acids are often called fatty acids and the esters are called triglycerides. The most common fatty acids are:

- octadeca-9-enoic (*oleic*) acid, an unsaturated acid which occurs in most fats and in olive oil;
- octadecanoic (*stearic*) acid, a saturated acid which occurs in animal fats;
- octadeca-9,12-dienoic (*linoleic*) acid, an unsaturated compound which is the principal acid in many vegetable oils, such as soya bean and corn oil.

Notes

Whereas the sodium salts of long-chain carboxylic acids are soluble in water, the calcium salts are not. Hence, in hard-water areas, the calcium ions form an unsightly scum with soap. Detergents, however, are sodium salts of sulfonic acids, which form soluble calcium salts, so that no scum is formed when a detergent is used:

O
R–⬡–S–O⁻ Na⁺
O

R = unbranched alkyl chain of about 12 carbon atoms

When fats are boiled with sodium hydroxide, glycerol is formed together with a mixture of the sodium salts of the three component acids. These salts are soaps:

$CH_2OOC(CH_2)_{16}CH_3$
|
$CHOOC(CH_2)_{16}CH_3$ + 3NaOH → $CHOH$ + $3CH_3(CH_2)_{16}COONa$
|
$CH_2OOC(CH_2)_{16}CH_3$

(products: CH_2OH / $CHOH$ / CH_2OH = propane-1,2,3-triol; $3CH_3(CH_2)_{16}COONa$ = sodium octadecanoate (a soap))

In practice, most soaps are mixtures of salts of long-chain carboxylic acids, since the fats and oils from which they are formed contain mixtures of these acids.

Glycerol, with its three O—H bonds, forms hydrogen bonds very easily and so has many applications that depend on its ability to attract water; it is used in the cosmetics industry, in food and in glues (to prevent materials drying too quickly). Glycerol is also an important component of wine, adding both to the sweetness and to the viscosity.

Biodiesel

Biodiesel is a renewable, non-petroleum-based fuel obtained mainly from vegetable oils by acid- or base-catalysed **transesterification**. Soya bean and rapeseed oils are most commonly used as biodiesel feedstocks. Waste vegetable oils and discarded animal fats are also used, but to a much lesser extent. In 2013, rapeseed oil accounted for 58% of total biodiesel production in the EU.

The most common form of biodiesel is a mixture of methyl esters of long-chain fatty acids, such as that formed with propane-1,2,3-triol in the following reaction:

$$\begin{array}{lcccccc} CH_2OOCR^1 & & & & CH_2OH & & R^1COOCH_3 \\ | & & & & | & & \\ CHOOCR^2 & + & 3CH_3OH & \rightleftharpoons & CHOH & + & R^2COOCH_3 \\ | & & & & | & & \\ CH_2OOCR^3 & & & & CH_2OH & & R^3COOCH_3 \end{array}$$

Biodiesel can be used alone (B100) or blended with petrodiesel. A fuel that can be used in unmodified diesel engines, containing 20% biodiesel and 80% petrodiesel, is labelled B20.

Biodiesel is non-toxic and **biodegradable**. It is cleaner burning than petrodiesel. New uses for the glycerol by-product are being explored. After several years of rapid growth, biodiesel production reached 10 million tonnes in Europe in 2011, since when there has been a small decline in both production and consumption.

There is a danger that farmers might stop planting staple food crops in favour of those that produce biofuels, leading to a shortage of food.

Notes

Transesterification is a reversible reaction in which an ester reacts with an alcohol, usually in excess, to form a new ester and a new alcohol.

Note that polyesters can be manufactured by transesterification (see this book, section 3.3.12.1).

3.3.9.2 Acylation

Acyl chlorides and acid anhydrides are useful synthetic intermediates for the preparation of other compounds. Both of these carboxylic acid derivatives possess good leaving groups (Cl^- and ^-OCOR, respectively) which also activate the adjacent carbonyl group by electron-withdrawal. Consequently, the carbon atom of the carbonyl group is susceptible to nucleophilic attack by water (hydrolysis) to give the acid, by alcohols to form esters, by ammonia to obtain amides and by amines to produce *N*-substituted amides (Fig 13). Although less vigorous, reactions of acid anhydrides with nucleophiles are analogous to those of acyl chlorides. Except in the case of hydrolysis, use of acid anhydrides results in the formation of co-products which are not easily removed.

Essential Notes

Note that a **good leaving group** is a stable species which is liberated in a chemical reaction; such groups are weak electron-pair donors (weak bases), and are readily replaced by better electron-pair donors (stronger bases).

$$\begin{array}{rcl} RCOCl + H_2O & \rightarrow & RCOOH + HCl \\ (RCO)_2O + H_2O & \rightarrow & 2RCOOH \\ & & \text{acid} \\ RCOCl + CH_3OH & \rightarrow & RCOOCH_3 + HCl \\ (RCO)_2O + CH_3OH & \rightarrow & RCOOCH_3 + RCOOH \\ & & \text{ester} \\ RCOCl + 2NH_3 & \rightarrow & RCONH_2 + NH_4Cl \\ (RCO)_2O + 2NH_3 & \rightarrow & RCONH_2 + RCOO^-NH_4^+ \\ & & \text{amide} \\ RCOCl + 2CH_3NH_2 & \rightarrow & RCONHCH_3 + CH_3NH_3^+Cl^- \\ (RCO)_2O + 2CH_3NH_2 & \rightarrow & RCONHCH_3 + RCOO^-CH_3NH_3^+ \\ & & N\text{-substituted amide} \end{array}$$

Fig 13
Acylation reactions

Notes

Acyl chlorides are reactive derivatives of carboxylic acids and can be obtained from the acids by reaction with sulfur dichloride oxide (*thionyl chloride*), as shown for ethanoyl chloride:

$$CH_3COOH + SOCl_2 \rightarrow CH_3COCl + SO_2 + HCl$$

Notes

Acid anhydrides are produced when acyl chlorides react with carboxylic acid salts:

$$RCOONa + RCOCl \rightarrow (RCO)_2O + NaCl$$

The mechanism of each of the above reactions involves nucleophilic addition to the carbonyl group, followed by elimination, and can be illustrated by specific examples (Fig 14 to Fig 17).

Fig 14
Hydrolysis of ethanoyl chloride

Fig 15
Formation of ethyl ethanoate

Fig 16
Formation of ethanamide (*acetamide*)

Fig 17
Formation of *N*-phenylethanamide (*acetanilide*)

The reaction between acyl chlorides and alcohols (Fig 15) is a highly effective way of producing esters. A base is usually added to neutralise the liberated hydrochloric acid. This method of preparation then avoids the equilibrium problem encountered in acid-catalysed esterification (see this book, section 3.3.9.1). In the case of amide formation (Fig 16), an excess of ammonia is used to neutralise the liberated hydrochloric acid. Acylation is used to form derivatives of both aliphatic and aromatic amines (Fig 17).

Industrial advantages of ethanoic anhydride

Unlike ethanoyl chloride, ethanoic anhydride is manufactured on a large scale for use as an acylating agent. The acid anhydride is relatively cheap compared with the acid chloride and is also less corrosive, less vulnerable to hydrolysis and less dangerous to use.

Uses of acylation reactions

A major use for ethanoic anhydride is in the manufacture of 2-ethanoyloxybenzenecarboxylic acid (aspirin). Aspirin is probably the most widely used drug of all time, being mainly employed as an

Notes

The preparation of a pure organic solid and testing its purity, and the preparation of a pure organic liquid are required practical activities.

Essential Notes

Note that the mechanisms of reactions involving ethanoic anhydride do not form part of the AQA A-level specification.

Notes

Unlike the reaction between ammonia and halogeno alkanes, further acylation is difficult because the lone pair on the amide nitrogen atom is withdrawn by the carbonyl group, so that the amide is less likely to act as a nucleophile:

$$R{-}C({=}O){-}NH_2 \longleftrightarrow R{-}C(O^-){=}\overset{+}{N}H_2$$

Consequently, amides are much less basic than amines (see this book, section 3.3.11.2).

Notes

Note that the mechanism of this type of acylation differs from that discussed in this book, section 3.3.10.2.

Essential Notes

Note that hydroxybenzene derivatives are called phenols.

analgesic (pain killer). The phenolic hydroxy group in the starting material, 2-hydroxybenzenecarboxylic (*salicylic*) acid, is acylated by heating the salicylic acid with ethanoic anhydride (Fig 18).

COOH, OH (ring) + $(CH_3CO)_2O$ ⟶ COOH, $OCOCH_3$ (ring) + CH_3COOH

Fig 18
Synthesis of 2-ethanoyloxybenzenecarboxylic acid (*aspirin*)

Another major use of ethanoic anhydride is for the conversion of cellulose into cellulose acetate (this is the preferred IUPAC name) for use as a synthetic fibre and in photographic film.

3.3.10 Aromatic chemistry

Benzene, C_6H_6, is the parent of a group of cyclic unsaturated compounds known as **arenes**. Traditionally, such compounds are referred to as being **aromatic** because many naturally-occurring fragrant substances were found to contain substituted benzene rings.

3.3.10.1 Bonding

The first satisfactory formula for benzene was put forward in 1865 by Kekulé, who suggested that the molecule had a cyclic arrangement of carbon atoms joined together by alternate single and double bonds. Two equivalent hexagonal structures can be drawn, so that isomeric 1,2-disubstituted benzenes might be expected:

X, X (ring) and X, X (ring)

In fact, only a single 1,2-disubstituted benzene exists. This observation was explained by Kekulé, who proposed that a rapid equilibrium existed between the two equivalent structures, thereby averaging out the single and double bonds.

Single-crystal X-ray diffraction analysis has shown that the benzene molecule is a planar, regular hexagon in which the carbon-carbon bond lengths are all the same (139 pm), being intermediate between normal single bonds (154 pm) and double (134 pm) bonds. The identical bonding between carbon atoms in benzene is implied by the use of a circle inside the hexagon to indicate six **delocalised** electrons:

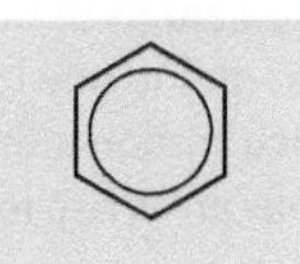

Notes

Benzene is often considered to be a **resonance hybrid** of the two Kekulé structures, neither of which actually exists:

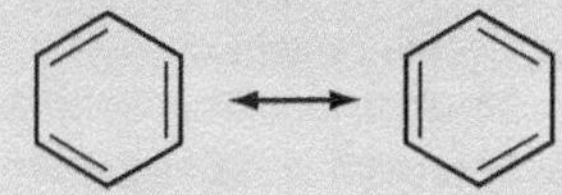

Notes

Care has to be taken when the circle notation is applied to polycyclic molecules, such as naphthalene:

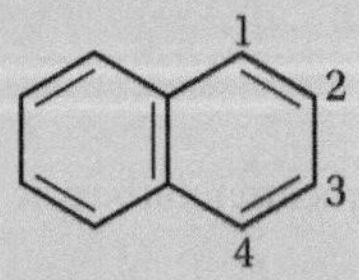

1, 2, 3, 4

This fused bicyclic arene has a total of **ten** p-electrons, not twelve, and should **not** be shown with circles in each of the two rings. It is not possible for both rings to be benzenoid at the same time and the 1,2-bond is shorter than the 2,3-bond.

Delocalisation stability

Thermochemical evidence shows that benzene is much more stable than the hypothetical cyclohexa-1,3,5-triene molecule would be. Thus, for example, the enthalpy change on hydrogenation of benzene is less exothermic than anticipated by comparison with cyclohexene (Fig 19).

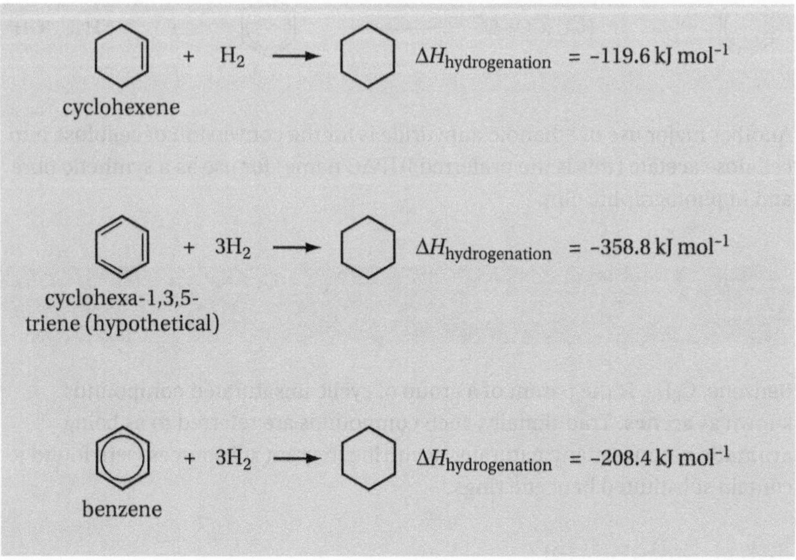

cyclohexene $+ H_2 \rightarrow$ cyclohexane $\Delta H_{hydrogenation} = -119.6\ kJ\ mol^{-1}$

cyclohexa-1,3,5-triene (hypothetical) $+ 3H_2 \rightarrow$ cyclohexane $\Delta H_{hydrogenation} = -358.8\ kJ\ mol^{-1}$

benzene $+ 3H_2 \rightarrow$ cyclohexane $\Delta H_{hydrogenation} = -208.4\ kJ\ mol^{-1}$

Fig 19
Enthalpies of hydrogenation

Assuming the enthalpy of hydrogenation of cyclohexa-1,3,5-triene to be three times that of cyclohexene gives the value of –358.8 kJ mol^{-1}. On this basis, benzene is 150.4 kJ mol^{-1} more stable than cyclohexa-1,3,5-triene. These differences are illustrated diagrammatically in Fig 20.

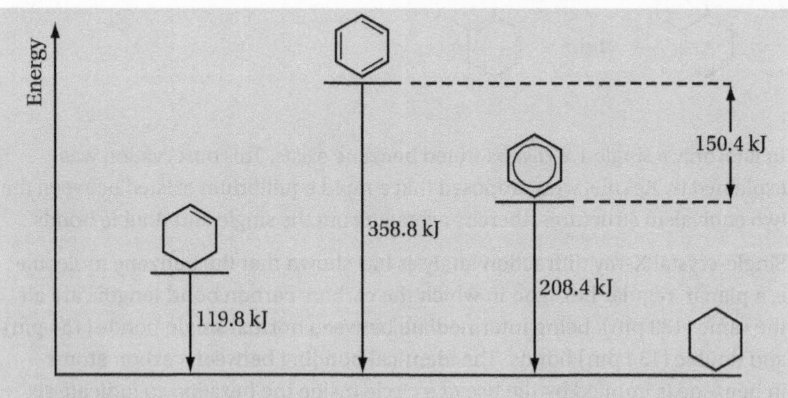

Fig 20
Hydrogenation of cyclohexene and benzene

Notes

It can be seen from this diagram that benzene is 150.4 kJ mol^{-1} **more** stable than the hypothetical molecule cyclohexa-1,3,5-triene.

All the bonds in benzene are identical because of the electronic structure of the molecule: each of the six carbon atoms in the hexagonal arrangement is bonded to two other carbon atoms and to one hydrogen atom, and all bond angles are 120°. Consequently, each carbon atom has one unused p-electron. Delocalisation of the six p-electrons gives rise to regions of electron density above and below the plane of the ring.

This cyclic electron delocalisation has a marked stabilising effect so that benzene undergoes overall addition reactions with difficulty. The increase in stability associated with electron delocalisation is called the **delocalisation energy** or the **resonance energy**.

3.3.10.2 Electrophilic substitution

The benzene molecule is susceptible to attack by positively-charged species: **electrophiles.** Although the first step in the substitution reaction involves addition of the electrophile (E^+), the resulting cationic intermediate then loses a proton to regenerate the delocalised (stable) system, resulting in the formation of a substitution product (Fig 21).

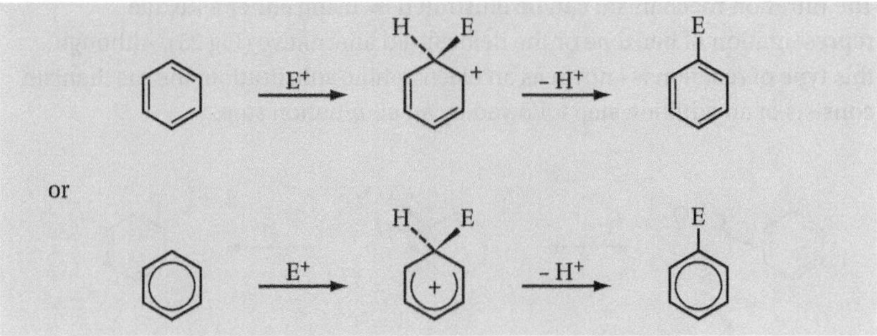

Essential Notes

Addition to benzene can occur, e.g. hydrogenation to form cyclohexane, but only with difficulty because the delocalisation energy is lost.

Notes

Note that 'wedges' and 'dotted' lines are used to indicate a 3D arrangement of atoms.

Fig 21
Electrophilic substitution of benzene

Electrophilic substitution reactions such as nitration and acylation are important steps in synthesis. In these reactions it is first necessary to form an electrophile, which then attacks the benzene ring.

Nitration

In nitration reactions, the electron-deficient **nitryl cation**, $^+NO_2$ (*nitronium ion*), is the electrophile. This species is generated via the protonation of nitric acid by a stronger acid, usually concentrated sulfuric acid. The overall equation:

$$HNO_3 + 2H_2SO_4 \rightleftharpoons {}^+NO_2 + H_3O^+ + 2HSO_4^-$$

summarises the three equilibria shown in Fig 22.

Essential Notes

Nitronium salts such as the tetrafluoroborate ($NO_2^+BF_4^-$) can also be used as nitrating agents.

$$HNO_3 + H_2SO_4 \rightleftharpoons [H_2NO_3]^+ + HSO_4^-$$

$$[H_2NO_3]^+ \rightleftharpoons H_2O + {}^+NO_2$$

$$H_2O + H_2SO_4 \rightleftharpoons H_3O^+ + HSO_4^-$$

Fig 22
Generation of the nitronium ion

Notes

Note that, in the first equilibrium, HNO_3 is acting as a base.

Breakdown of the protonated nitric acid molecule gives water and the nitronium ion (Fig 23).

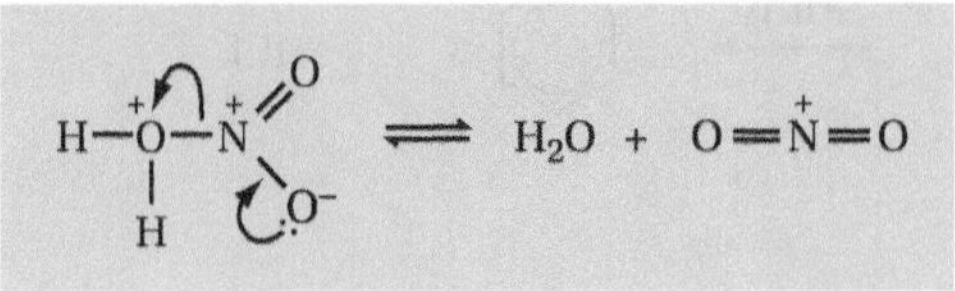

Fig 23
Formation of $^+NO_2$

Nitrobenzene is obtained by warming benzene with a mixture of concentrated nitric acid and concentrated sulfuric acid (Fig 24). The sulfuric acid aids the formation of the inorganic electrophile by the removal of water (Fig 22).

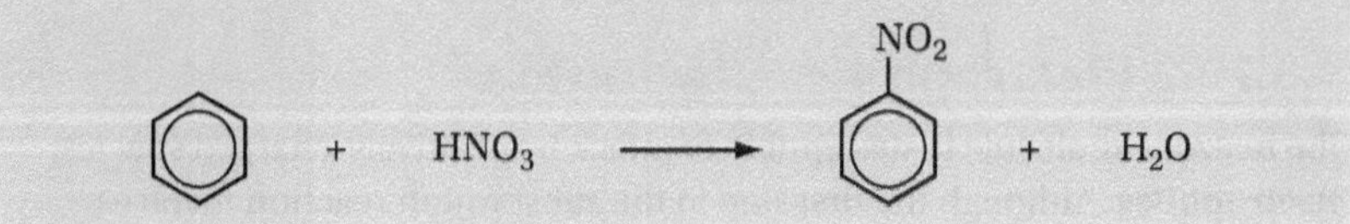

Fig 24
Nitration of benzene

The nitration mechanism can be illustrated by using either a Kekulé representation of benzene or the delocalised alternative (Fig 25). Although this type of reaction is known as an electrophilic substitution, the mechanism consists of an addition step followed by an elimination step.

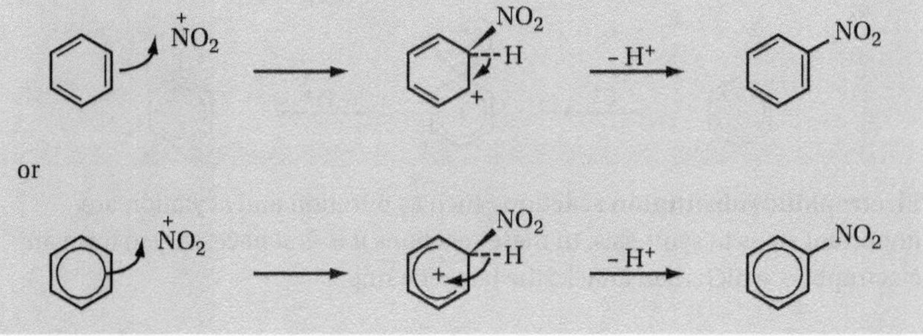

Fig 25
Mechanism of nitration

Use of nitration reactions

Aromatic nitro compounds are important as precursors of aromatic amines. The reduction of nitro compounds can be achieved easily by catalytic hydrogenation (e.g. Ni/H_2) or by using metal/acid combinations (e.g. Sn/HCl) (see this book, section 3.3.11.1). The resulting primary aromatic amines are very useful intermediates in organic synthesis.

Friedel-Crafts acylation reactions

Friedel-Crafts reactions are important examples of electrophilic substitution because they lead to carbon-carbon bond formation. In 1877, Friedel and Crafts discovered that a halogenoalkane will react with benzene in the presence of a catalyst such as aluminium chloride (Fig 26).

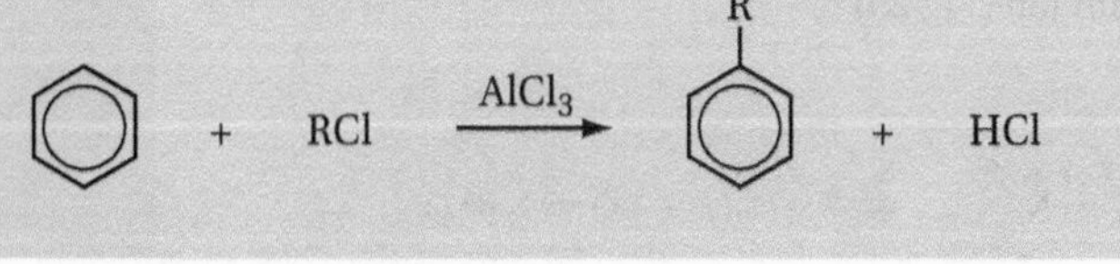

Fig 26
Friedel-Crafts alkylation of benzene

Notes

Nitration of methylbenzene is easier than nitration of benzene due to activation by the electron-releasing methyl group, which increases the electron density in the benzene ring. However, the presence of electron-withdrawing nitro groups makes each successive substitution progressively slower. In the case of benzene itself, the rate of formation of nitrobenzene from benzene is about 10^7 times faster than the rate of formation of 1,3-dinitrobenzene from nitrobenzene.

Notes

The positive charge in the intermediate cation can be delocalised by means of electron shifts involving the two remaining conjugated double bonds.

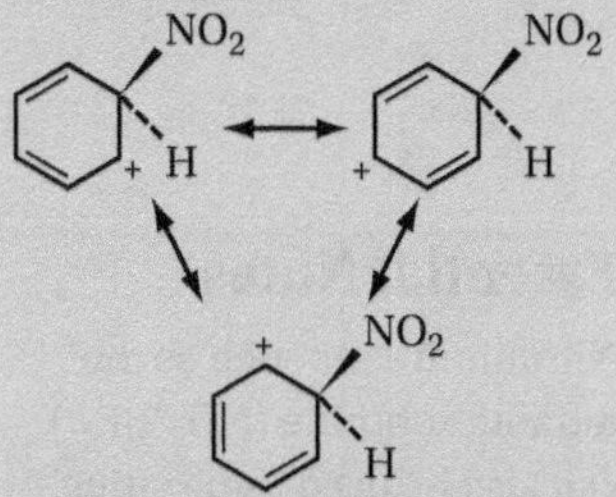

The overall picture is well represented by the structure derived from the delocalised benzene ring:

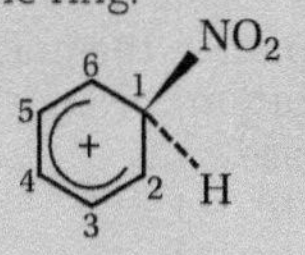

Note that carbon 1 is saturated and tetrahedral.

Notes

Note that either of these representations of electrophilic substitution is acceptable in examination answers.

Alkylation reactions do not constitute important steps in synthesis, unlike corresponding acylation reactions which introduce a reactive functional group into the aromatic ring. Thus, ketones are produced when benzene reacts with acyl chlorides, in the presence of $AlCl_3$. The **acylium cation** initially formed is sufficiently electrophilic to attack benzene by the usual aromatic substitution mechanism (Fig 27).

Notes
Only Friedel–Crafts acylation reactions are included in the AQA A-level specification.

$$RCOCl + AlCl_3 \rightarrow RCO^+ + AlCl_4^-$$

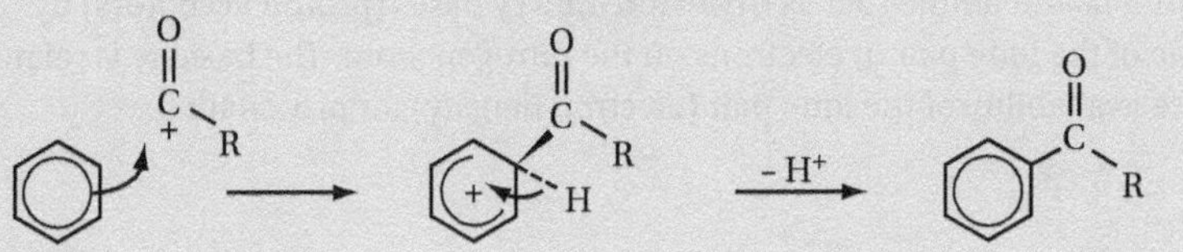

Fig 27
Mechanism of acylation

Notes
Note that the mechanism of this type of acylation differs from that discussed in this book, section 3.3.9.2

The formation of phenylethanone provides a specific example:

$$C_6H_6 + CH_3COCl \rightarrow C_6H_5COCH_3 + HCl$$

Carboxylic acid anhydrides are sometimes used instead of acyl chlorides. The acylium ion is formed in the presence of aluminium chloride:

$$(RCO)_2O + AlCl_3 \rightarrow RCO^+ + RCOOAlCl_3^-$$

3.3.11 Amines

Replacement of the hydrogen atoms in ammonia by alkyl or aryl groups gives rise to three types of amine:

RNH_2	R_2NH	R_3N
primary	secondary	tertiary

The most important are primary amines, e.g. CH_3NH_2 (aliphatic) or $C_6H_5NH_2$ (aromatic). Boiling points are lower than those of the analogous alcohols, e.g. CH_3NH_2 is gaseous, because of weaker hydrogen bonding.

3.3.11.1 Preparation

A useful general method for preparing primary aliphatic amines of the type RCH_2NH_2 involves a two-step synthesis with a halogenoalkane as the starting material. Nucleophilic substitution with cyanide ion in aqueous ethanol gives the corresponding nitrile, which can then be reduced to a primary amine:

Notes
Note that $NaBH_4$ is not sufficiently powerful to reduce a CN group.

$$RBr + CN^- \rightarrow RC{\equiv}N + Br^-$$

$$RC{\equiv}N + 2H_2 \rightarrow RCH_2NH_2$$

Catalytic hydrogenation (e.g. Ni/H_2) is often used for the reduction as is lithium tetrahydridoaluminate(III) ($LiAlH_4$) in dry ethoxyethane.

Essential Notes
Just as R is often used to represent any alkyl group, Ar is similarly used to indicate an aryl group.

Primary aromatic amines are usually prepared by the reduction of nitro compounds:

$$ArNO_2 + 3H_2 \xrightarrow{Ni} ArNH_2 + 2H_2O$$

Notes

Many functional groups can be introduced into benzene via the conversion of $ArNH_2$ into ArN_2^+ in a process called **diazotisation**. This process does not form part of the AQA A-level specification.

Again, catalytic hydrogenation is a suitable method, giving almost quantitative (100%) yields. Methods of chemical reduction include heating with Sn/HCl (laboratory) or with Fe/HCl (industry). In these cases, the resulting amine is present in solution as $[ArNH_3]^+$. Consequently, a base (e.g. NaOH) is added to liberate the free amine. The organic product is then removed from the reaction mixture by distillation.

3.3.11.2 Base properties

Notes

K_a values for bases refer to dissociation of the conjugate acid:

$RNH_3^+ \rightleftharpoons RNH_2 + H^+$

so that the larger the K_a value (equilibrium to the right), the smaller the pK_a value and the weaker the base.

Ammonia and amines act as **Brønsted–Lowry** bases (proton acceptors) by virtue of the lone pair of electrons on the nitrogen atom. The basicity is related to the availability of the lone pair (electron density) for protonation:

$$R\ddot{N}H_2 \quad H^+ \rightleftharpoons RNH_3^+$$

The introduction of one alkyl group into ammonia strengthens the base due to the inductive effect of the alkyl group, which pushes electrons towards the nitrogen atom. For example, methylamine (pK_a 10.6) is a stronger base than ammonia (pK_a 9.2). On the other hand, arylamines are significantly weaker bases than alkylamines because of involvement of the lone pair in the aromatic delocalisation, leading to a decrease in electron density on the nitrogen atom. Thus, for example, phenylamine (pK_a 4.6) is a weaker base than ammonia.

3.3.11.3 Nucleophilic properties

Essential Notes

Ligands are also lone-pair donors (see *Collins Student Support Materials: A-Level year 2 - Inorganic and Relevant Physical Chemistry*, section 3.2.5.1).

Ammonia and amines also act as nucleophiles (electron-pair donors) and take part in nucleophilic substitution reactions. The reaction between a halogenoalkane and ammonia is an alkylation producing an alkylammonium salt. Proton exchange with another ammonia molecule produces the primary amine:

$$NH_3 + RBr \rightarrow [RNH_3]^+Br^-$$

$$[RNH_3]^+Br^- + NH_3 \rightleftharpoons \underset{\text{primary}}{RNH_2} + [NH_4]^+Br^-$$

Further substitution is possible, because the primary amine can compete effectively with ammonia, as the nucleophile, for reaction with the halogenoalkane to generate a dialkylammonium salt. Further proton exchange with either ammonia or with RNH_2 liberates the secondary amine:

$$RNH_2 + RBr \rightarrow [R_2NH_2]^+Br^-$$

$$[R_2NH_2]^+Br^- + NH_3 \rightleftharpoons \underset{\text{secondary}}{R_2NH} + [NH_4]^+Br^-$$

A third alkylation can then take place to give a trialkylammonium salt which, in turn, will donate a proton to ammonia or to another amine:

$$R_2NH + RBr \rightarrow [R_3NH]^+Br^-$$

$$[R_3NH]^+Br^- + NH_3 \rightleftharpoons \underset{\text{tertiary}}{R_3N} + [NH_4]^+Br^-$$

Essential Notes

Note that a quaternary ammonium ion with four different groups is chiral.

The resulting tertiary amine then reacts with the halogenoalkane in a fourth alkylation step to form a quaternary ammonium salt:

$$R_3N + RBr \rightarrow \underset{\text{quaternary ammonium salt}}{[R_4N]^+Br^-}$$

A mixture of products is usually obtained. Clearly, this outcome limits the usefulness of direct alkylation in synthesis, although separation of the various components is possible. A high yield of the quaternary ammonium salt is obtained by using a large excess of halogenoalkane. On the other hand, a large excess of ammonia reduces the possibility of further substitution and gives a better yield of primary amine.

Each of the four steps in the above sequence is mechanistically similar and involves nucleophilic attack from the side opposite the leaving group, as illustrated in Fig 28.

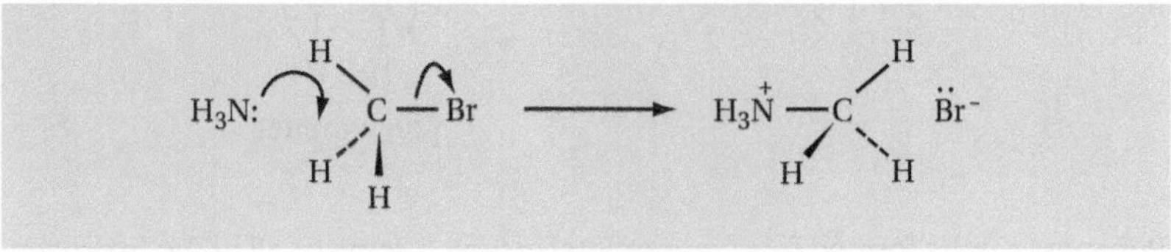

Fig 28
Formation of methylammonium bromide

> **Notes**
> **Surfactants** (derived from surface acting agents) are wetting agents that lower the surface tension of a liquid and the interfacial tension between two liquids. Such chemicals contain both hydrophobic and hydrophilic groups and are soluble both in organic solvents and in water.

Quaternary ammonium salts possessing two long-chain alkyl groups, such as $[CH_3(CH_2)_{17}]_2N(CH_3)_2^+Cl^-$, are used as cationic surfactants in fabric softening.

3.3.12 Polymers

Polymers (meaning many units) are effectively long chains of molecules which are usually organic or biological in nature, though some may be inorganic, e.g. silicone rubber.

The term polymer is often used as a synonym for plastic. However, whereas all plastics are polymers, not all polymers are plastic. Before conversion into plastic products, polymers are often modified by the incorporation of plasticisers, stabilisers and colourants.

> **Notes**
> The bond energies of C=C and C—C bonds are 612 and 348 kJ mol^{-1}, respectively. Polymerisation, which involves forming two C—C bonds from one C=C bond, is therefore an exothermic process.

Addition polymers

Alkene molecules link together in the presence of a catalyst to form addition polymers which are saturated, such as poly(ethene). A section of the polymer (formed from eight ethene molecules) is shown in Fig 29.

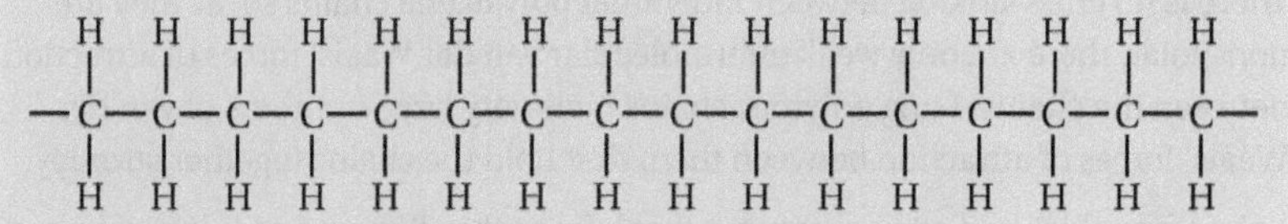

Fig 29
Structure of poly(ethene)

Addition polymers are also known as **chain-growth** polymers. They are formed by the addition of monomers to the end of a growing chain. The end of the chain is reactive because it is a radical, which is formed at the beginning of the reaction by use of catalysts such as organic peroxides, i.e. ROOR. Peroxide molecules readily split into radicals to initiate the chain growth. A radical re-forms at the end of the chain after each addition of a monomer molecule.

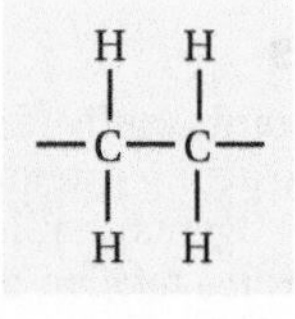

$$-\overset{\displaystyle H}{\underset{\displaystyle H}{\overset{|}{\underset{|}{C}}}}-\overset{\displaystyle H}{\underset{\displaystyle H}{\overset{|}{\underset{|}{C}}}}-$$

Fig 30
The repeating unit of poly(ethene)

It is usual to ignore any consideration of the end-groups that come from the catalyst, as these represent an insignificant fraction of a large polymer.

Polymers formed from alkenes are usually represented using a **repeating unit**, such as that for ethene shown in Fig 30.

The polymerisation of ethene can therefore be represented by the equation in Fig 31.

Fig 31
The formation of poly(ethene)

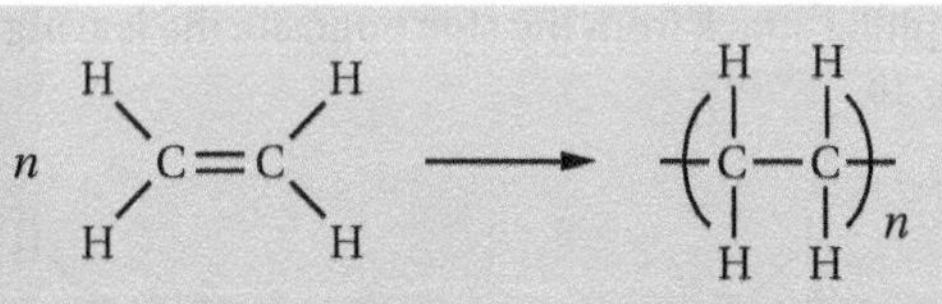

$$n\,H_2C{=}CH_2 \longrightarrow \left(\!-CH_2-CH_2-\!\right)_n$$

poly(ethene)

Notes

n represents a large whole number, which is the number of individual molecules (monomers) joining together to form the polymer.

Polymers can be formed from monomers, in which some or all of the hydrogen atoms in ethene have been replaced. One example of this, poly(phenylethene) *(polystyrene)* is considered in this section (see Fig 32).

Fig 32
Formation of poly(phenylethene) *(polystyrene)*

$$n\,H_2C{=}CH(C_6H_5) \longrightarrow \left(\!-CH_2-CH(C_6H_5)-\!\right)_n$$

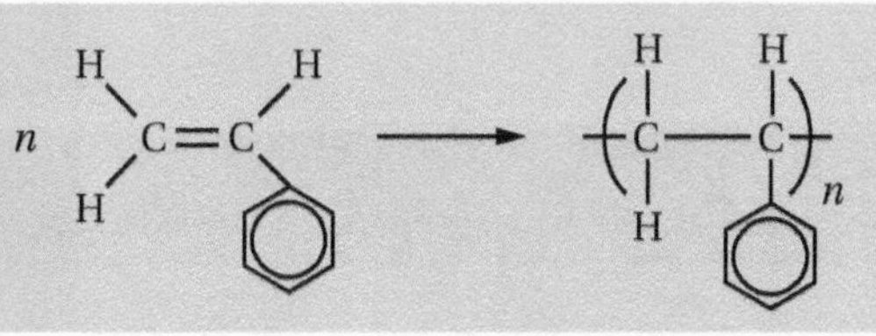

IUPAC rules for the naming of addition polymers use the monomer name in brackets, preceded by 'poly'. For example, ethene polymerisation produces the polymer poly(ethene), and propene polymerisation produces poly(propene). Using this system, the polymer from chloroethene is called poly(chloroethene), but its non-IUPAC name, poly(vinyl chloride) (PVC) is still widely used.

Unlike alkenes, polyalkenes are saturated and therefore are unreactive, like simple alkanes, owing to strong covalent bonds between atoms and a lack of bond polarity. Polyalkenes are also non-biodegradable. However, polymers of this kind are, nevertheless, highly flammable. Methods of disposal of polymers are considered in this book, section 3.3.12.2.

There is no cross-linking between individual polyalkene chains so, as they are non-polar, there are only weak intermolecular van der Waals' forces of attraction between the chains. Long polymer chains have very large numbers of van der Waals' forces of attraction between them that hold the chains together strongly.

Poly(ethene) is used as packaging, especially as thin film and as 'plastic' bags.

Poly(propene) is very versatile. It is used to make rigid containers and objects such as car bumpers. It can also be made into fibres which are used as the backing for carpets and in thermal clothing.

Poly(chloroethene) (PVC) has a range of uses. Examples are window frames, guttering, plumbing and leather-look fabrics. A **plasticiser** is often incorporated to soften it and to make it more flexible, but there have been environmental concerns about some of these. For this reason, cling film (for food use) is increasingly made from poly(ethene) rather than from PVC with added plasticiser.

Some polymers, such as poly(propene), can be recycled. Containers made of poly(propene) are collected, cleaned and cut into small pieces. The plastic is then melted and remoulded into a new object or extruded and spun into fibres.

Over time, knowledge and understanding of the production and properties of polymers has developed, leading to an increased ability to control closely the length of polymer chains. Our ability to tailor the properties of polymers for specific uses and applications in industry has increased with the use of co-polymers (made by using two different monomers).

Notes

At low temperatures, long-chain polymers are hard and glass-like. As the temperature is raised, the polymer passes through a **glass-transition temperature** (T_g) after which it becomes more flexible and mouldable (plastic); such polymers are said to be thermoplastic. In order to make synthetic polymers more flexible, non-volatile liquids (plasticisers) are added which soften the polymer and lower T_g (see this book, section 3.3.9.1). For example, the value of T_g for PVC can be lowered from 80 °C to about 0 °C by the addition of dibutyl benzene-1,2-dicarboxylate.

3.3.12.1 Condensation polymers

Condensation polymers (or **step-growth** polymers) are formed by the reaction between molecules having two functional groups, involving the loss of small molecules such as H_2O, CH_3OH or HCl.

Examples of condensation polymers include:

- polyesters, formed by reactions between dicarboxylic acids and diols
- polyamides, formed by reactions between dicarboxylic acids and diamines
- **proteins**, formed by reactions between amino acids (see this book, section 3.3.13.2).

Polyesters

The reaction between a dicarboxylic acid and a diol leads to the formation of a polyester. The most important polyester is **Terylene** which can be formed from benzene-1,4-dicarboxylic (*terephthalic*) acid and ethane-1,2-diol (Fig 33).

A better product is obtained by a transesterification (see this book, section 3.3.9.1) reaction between the dimethyl ester of the acid and the diol. In this case, methanol is evolved as a gas.

Polyesters, just like single esters (see this book, section 3.3.9.1), are susceptible to hydrolysis and can be broken down into the component monomers. Consequently, polyesters, like polyamides, are biodegradable polymers, as shown in Fig 34.

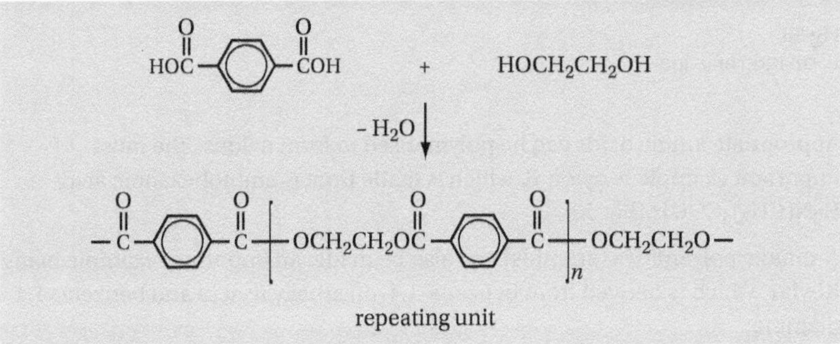

$$HOOC{-}C_6H_4{-}COOH + HOCH_2CH_2OH \xrightarrow{-H_2O} {-}CO{-}C_6H_4{-}\left[CO{-}OCH_2CH_2OCO{-}C_6H_4{-}CO\right]_n{-}OCH_2CH_2O{-}$$

repeating unit

Fig 33
Formation of Terylene

Essential Notes

Molten polymers can be spun into a fibre or cast into a film. Typical uses include permanent-press fabrics and magnetic recording tapes. Terylene is also blow-moulded on a large scale to make plastic bottles.

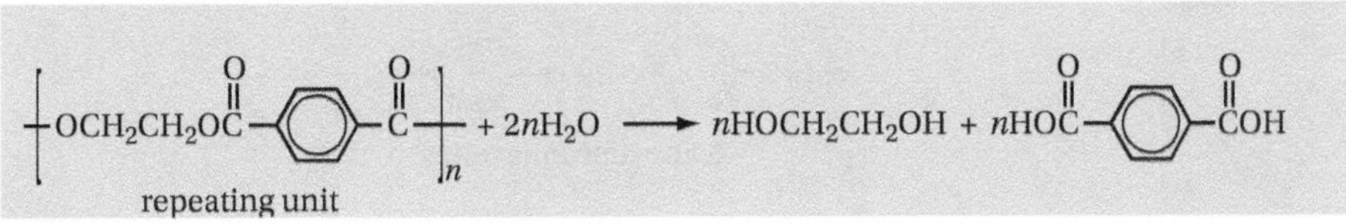

$$\left[OCH_2CH_2OCO{-}C_6H_4{-}CO\right]_n + 2nH_2O \longrightarrow nHOCH_2CH_2OH + nHOOC{-}C_6H_4{-}COOH$$

repeating unit

Fig 34
Hydrolysis of the polyester Terylene

Essential Notes

In the laboratory, nylon-6,6 can be pulled as a filament or rope from the interface between an aqueous solution of hexane-1,6-diamine and a solution of hexanedioyl dichloride in a solvent, e.g. hexane, which is immiscible with water.

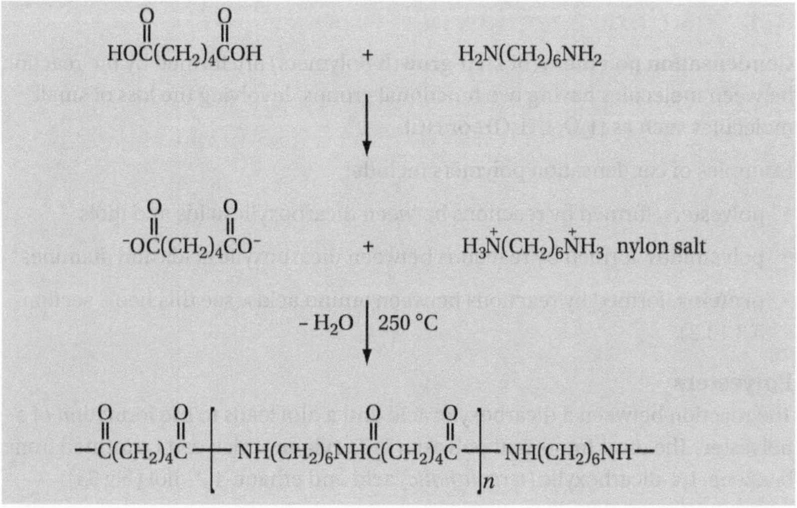

Polyamides

The reaction between a dicarboxylic acid and a diamine leads to the formation of a polyamide. Nylon-6,6 is formed from hexanedioic acid and hexane-1,6-diamine (Fig 35); the repeating unit is enclosed in brackets.

Nylon-6,6 is so called because it is derived from a six-carbon diacid and a six-carbon diamine.

Fig 35
Formation of nylon-6,6

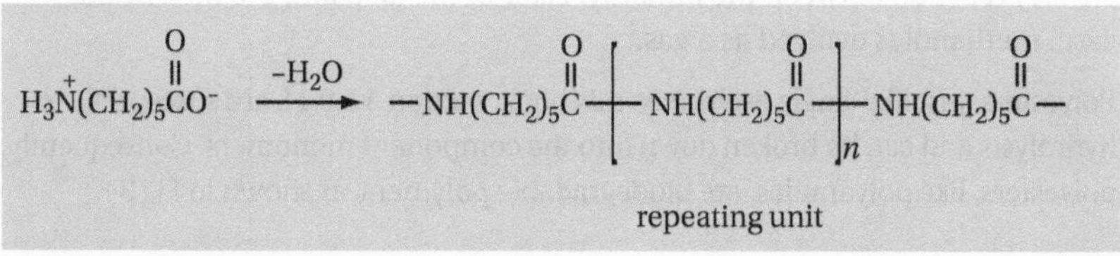

$$HOOC(CH_2)_4COOH + H_2N(CH_2)_6NH_2$$

$$\downarrow$$

$$^-OOC(CH_2)_4COO^- + H_3\overset{+}{N}(CH_2)_6\overset{+}{N}H_3 \quad \text{nylon salt}$$

$$\downarrow -H_2O \quad 250\,°C$$

$$-C(=O)(CH_2)_4C(=O)-\left[NH(CH_2)_6NHC(=O)(CH_2)_4C(=O)\right]_n-NH(CH_2)_6NH-$$

$$H_3\overset{+}{N}(CH_2)_5COO^- \xrightarrow{-H_2O} -NH(CH_2)_5C(=O)-\left[NH(CH_2)_5C(=O)\right]_n-NH(CH_2)_5C(=O)-$$

repeating unit

Fig 36
Formation of nylon-6

Notes

Nylon first found widespread use in textiles and carpets. Because of its resistance to stress, other uses include tyre cords, fishing lines, mountaineering ropes and bearings and gears.

Appropriate amino acids can be polymerised to form nylons. The most important example is nylon-6, which is made from 6-aminohexanoic acid, $H_2N(CH_2)_5COOH$ (Fig 36).

Aromatic polyamides (aramids) can also be made, an important example being **Kevlar**, which is derived from benzene-1,4-dicarboxylic acid and benzene-1,4-diamine:

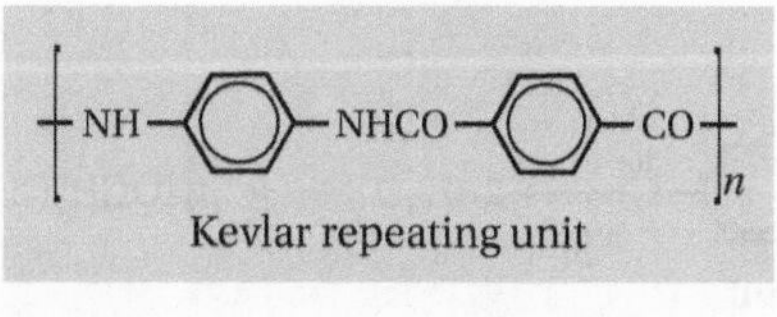

$$\left[-NH-C_6H_4-NHCO-C_6H_4-CO-\right]_n$$

Kevlar repeating unit

Hydrogen bonding between polymer chains results in a tough, sheet-like structure:

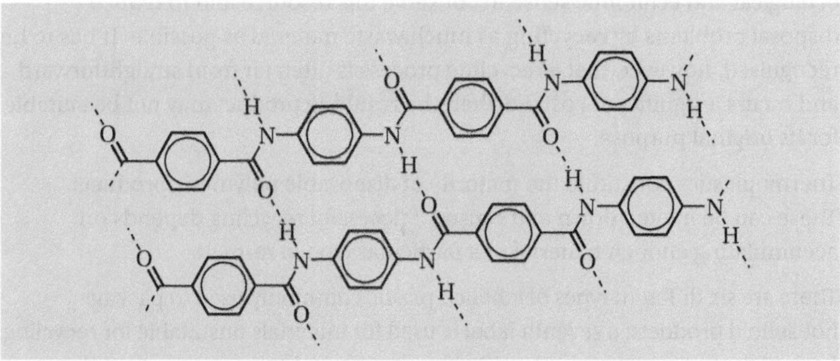

Kevlar has a tensile strength greater than that of steel and is a component of some bullet-proof vests. Other uses include the protective outer sheath of fibre optic cable and the inner lining of some bicycle tyres.

Essential Notes

Nomex, the 1,3-linked isomer of Kevlar, is used in flame-resistant clothing for firefighters, military pilots and racing car drivers.

3.3.12.2 Biodegradability and disposal of polymers

Most common addition polymers are not biodegradable (they cannot be broken down by micro-organisms). Polyalkenes contain no functional groups and so are chemically inert and resistant to biodegradation. Although other addition polymers contain functional groups, the materials they form are not biodegradable. Condensation polymers, typically polyamides and polyesters, possess repeating units which can be split by hydrolysis; such polymers are biodegradable by enzyme action.

For years, most of the plastic waste in the UK and elsewhere has gone into **landfill sites** (rubbish dumps), many of which have now reached capacity. Plastic waste accounts for only approximately 8% of the mass, but makes up about 20% of the volume of the rubbish.

The incorporation of cellulose, starch or protein into synthetic polymers or plastic articles makes them biodegradable, so that disintegration can occur in landfill sites. It is now possible to design new materials to be completely biodegradable. However, traditional synthetic plastics are currently more economically attractive than biodegradable ones. It is likely that the need for more environmentally responsible disposal will gradually lead to an increased availability of biodegradable materials.

Notes

Ecoflex is a fully biodegradable aliphatic–aromatic co-polyester based on butane-1,4-diol, hexanedioic acid and benzene-1,4-dicarboxylic acid. The material is used in film form for disposable packaging and will decompose within a few weeks when composted.

Incineration

Polymers can be burned to recover a significant amount of energy for power generation or heating. The volume of waste is greatly reduced by such **incineration,** but the process may also generate toxic gases.

Modern incinerator designs include a high-temperature zone, where the flue gases encounter a temperature in excess of 850 °C for at least 2 seconds to ensure the complete breakdown of any organic toxins. Incineration of household waste reduces its volume by about 95%.

Before dispersal into the atmosphere, flue gases are cleaned to extract pollutants such as SO_2, NO_x and HCl. On average, about 50% of the chlorine input into municipal incinerators comes from PVC.

Recycling

The prime source of nearly all polymers is non-renewable crude oil. It makes ecological and economic sense to conserve this resource and to reduce disposal problems by **recycling** as much waste material as possible. It has to be recognised, however, that a recycling process is often far from straightforward and incurs a significant cost. Further, the resulting product may not be suitable for its original purpose.

Thermoplastics constitute the majority of disposable polymeric products. These can be melted down and reused. Successful recycling depends on accumulating enough material of a particular type to re-melt.

Notes

The various categories of plastic listed here are not required knowledge and do not form part of the AQA A-level specification.

There are six different types of labelled plastic commonly used to package household products; a seventh label is used for materials unsuitable for recycling. Each of the various types is identified by a numerical code on the packaging:

- **Type 1: PETE** – *polyethylene terephthalate* (PET), or Terylene (poly(ethyl benzene-1,4-dicarboxylate))
  - often first used for water and beverage containers
  - recycled for use in textiles, e.g. fibres and carpets.
- **Type 2: HDPE** – *high-density polyethylene* (high-density polyethene)
  - often first used for opaque milk and motor-oil containers
  - recycled for use in coloured products, e.g. bottle crates, drainage pipes and stadium seats.
- **Type 3: V** – *polyvinyl chloride* (PVC) (poly(1-chloroethene))
  - used in cling film, vegetable oil bottles and alcoholic beverage containers
  - recycled products include drainage pipes, floor tiles, non-food bottles and protective bubble wrap.
- **Type 4: LDPE** – *low-density polyethylene* [less dense and more flexible than HDPE] (low-density polyethene)
  - widely used for many kinds of plastic bags, often coloured
  - recycled mainly as black garbage bags and compost bins.
- **Type 5: PP** – *polypropylene* (poly(propene))
  - used for relatively strong food containers, e.g. ketchup bottles and margarine tubs
  - recycled products, often coloured, include cafeteria trays, ice scrapers, and VCR storage cases, as well as car bumpers and under-body parts.
- **Type 6: PS** – *polystyrene* (poly(1-phenylethene))
  - used in its expanded form for protective packing and insulated utensils
  - recycling is difficult and, as an alternative, waste material is sometimes used as a filler in other plastics and concrete.
- **Type 7: Other** – covers mixed plastics which have no recycling potential.

Great care has to be taken during recycling not to mix the various types of plastic. The presence of just one rogue PVC bottle in a melt of 10,000 PET bottles will ruin the whole batch.

At the present time, the multiple recycling of paper, glass and metals (which all become products similar to their source materials) is carried out on a much larger scale than is used for plastics, so that recycling of plastics has only a minimal impact on the amount of natural resources and energy used.

3.3.13 Amino acids, proteins and DNA

3.3.13.1 Amino acids

The term **amino acid** is commonly used for compounds which have a primary amino group attached to the carbon atom adjacent to a carboxylic acid group:

$RCH(NH_2)COOH$

The position next to the carboxyl functional group is sometimes called the α-position, and a more complete name for such compounds is α-aminocarboxylic acids.

There are 20 naturally-occurring amino acids, which differ only in the nature of the group R. Apart from the simplest example, aminoethanoic acid, *glycine*, in which R = H, these compounds show optical activity. However, only one of the enantiomers occurs naturally. The amino acids are usually called by their common or trivial names. Some examples are given in Fig 37.

Notes

Examples of amino acids are included in the AQA Data sheet.

$CH_3CH(NH_2)COOH$	$HOCH_2CH(NH_2)COOH$
alanine	serine
–$CH_2CH(NH_2)COOH$	$H_2NCOCH_2CH(NH_2)COOH$
phenylalanine	asparagine

Fig 37
Examples of α-aminocarboxylic acids

Acid and base properties

Every amino acid has a carboxyl group and an amino group. So, depending on the pH of the solution in which the compound is dissolved, each group can exist in an ionic or a non-ionic form. The carboxyl groups of amino acids have pK_a values of about 2, whereas the protonated amino groups have pK_a values close to 9. Thus, in a very acidic solution, the amino group is protonated and the carboxyl group is undissociated. Conversely, in a strongly basic solution, the carboxyl group is deprotonated and the amino group is unchanged. In neutral solution, the carboxyl group is deprotonated and the amino group is protonated (Fig 38).

Notes

The **isoelectric point** of an amino acid is the pH at which it has no net charge; for simple amino acids, this point is midway between the two pK_a values. In the case of alanine, for example, the pK_a values are approximately 2.3 and 9.7, respectively, leading to an isoelectric point at a pH value of 6.0.

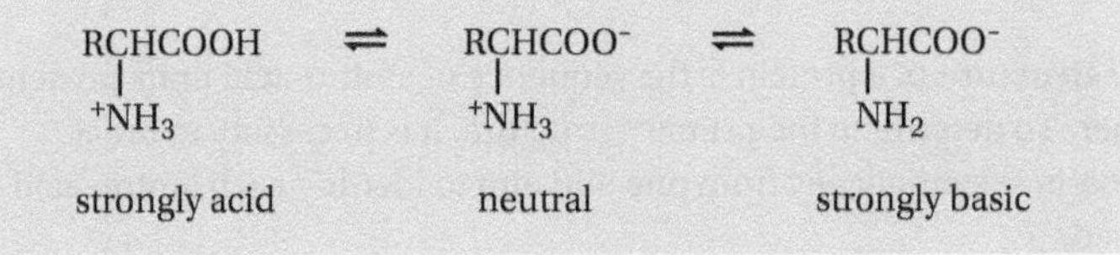

Fig 38
Amino acids in solution

It is important to realise that an amino acid can never exist as an uncharged compound, regardless of the pH of the solution. At a pH of 7.3, found in living systems, an amino acid exists as a dipolar ion, $RCH(NH_3^+)COO^-$, which has an overall neutral charge. Such a species is called a **zwitterion**.

The zwitterionic nature of amino acids is reflected in the relatively high melting points of the crystals. Thus, glycine itself melts at about 290 °C, whereas the corresponding hydroxyethanoic acid, $HOCH_2COOH$, which is extensively hydrogen bonded but not ionic, melts at 80 °C.

3.3.13.2 Proteins

Proteins are naturally-occurring polymers of amino acids joined together by amide bonds (Fig 39) (see this book, section 3.3.12.1).

Fig 39
Linking together amino acids

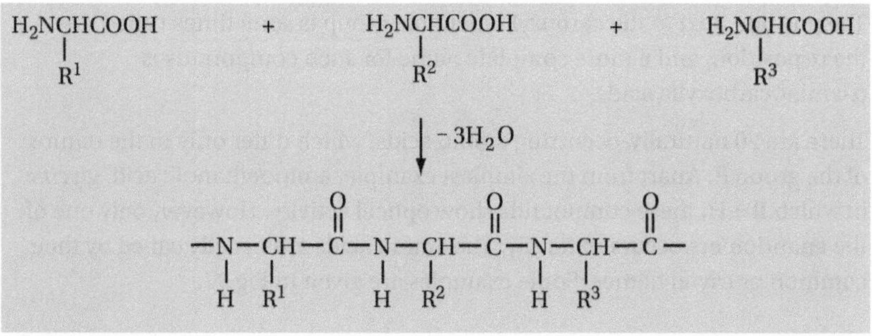

$$\underset{R^1}{H_2NCHCOOH} + \underset{R^2}{H_2NCHCOOH} + \underset{R^3}{H_2NCHCOOH} \xrightarrow{-3H_2O} -\underset{H}{N}-\underset{R^1}{CH}-\overset{O}{\overset{\|}{C}}-\underset{H}{N}-\underset{R^2}{CH}-\overset{O}{\overset{\|}{C}}-\underset{H}{N}-\underset{R^3}{CH}-\overset{O}{\overset{\|}{C}}-$$

Essential Notes

Proteins can be divided into two classes. **Fibrous** proteins contain long chains of polypeptides which occur in bundles (e.g. keratin); these proteins are insoluble in water. **Globular** proteins are folded into roughly spherical shapes and are soluble in water (e.g. haemoglobin).

The amide bond (CO–NH) in proteins is often called a **peptide link**. Peptide is the term used to describe sequences of relatively few amino acids. For example, a tripeptide is made from three amino-acid units. Proteins can be thought of as complex, naturally-occurring polypeptides, which are made up of about 40 to around 4000 amino-acid units.

Hydrolysis of the peptide link

Peptides and proteins can be hydrolysed to the constituent amino acids in the presence of a strong acid or by the action of a specific enzyme catalyst. Fission of the peptide bonds in a protein liberates the component amino acids (Fig 40).

Fig 40
Hydrolysis of peptide bonds

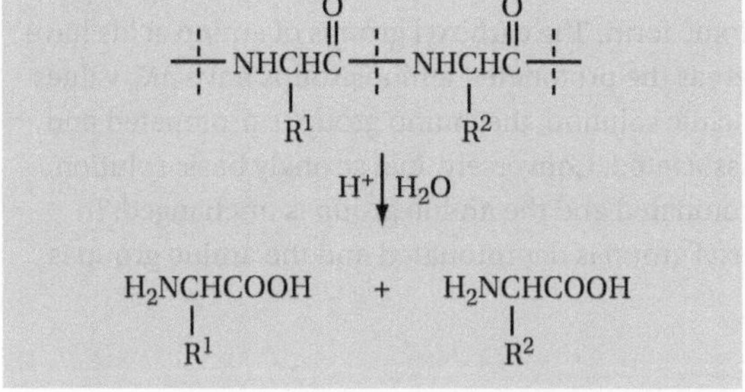

$$\Big|\,\underset{R^1}{NHCH}\overset{O}{\overset{\|}{C}}\,\Big|\,\underset{R^2}{NHCH}\overset{O}{\overset{\|}{C}}\,\Big| \xrightarrow{H^+,\ H_2O} \underset{R^1}{H_2NCHCOOH} + \underset{R^2}{H_2NCHCOOH}$$

The **primary structure** of a protein is the sequence of amino-acid units present in the polymer. To determine the primary structure, it is necessary to break down the protein systematically from one end and to identify each amino acid as it is released.

Mixtures of amino acids can be separated, and identified, by paper chromatography or by thin-layer chromatography. Developing agents, such as ninhydrin or ultraviolet light, can be used to locate them (see this book, section 3.3.16).

Hydrogen bonding in proteins

Peptide chains tend to form orderly, hydrogen-bonded arrangements. In particular, the carbonyl oxygen atoms form hydrogen bonds with the amide hydrogen atoms of other peptide groups, i.e. C=O---H–N. The hydrogen-bonded arrangements give rise to two distinct types of **secondary structure**, either an **α-helix** or a **β-pleated sheet.**

Despite hydrogen bonds being relatively weak, the behaviour of a large number of hydrogen bonds is cooperative, leading to the formation of stable structures. Most proteins have some helical structure. In hair and wool, the helices are coiled around each other to form rope-like structures.

Notes

These structures are not required by the AQA A-level specification.

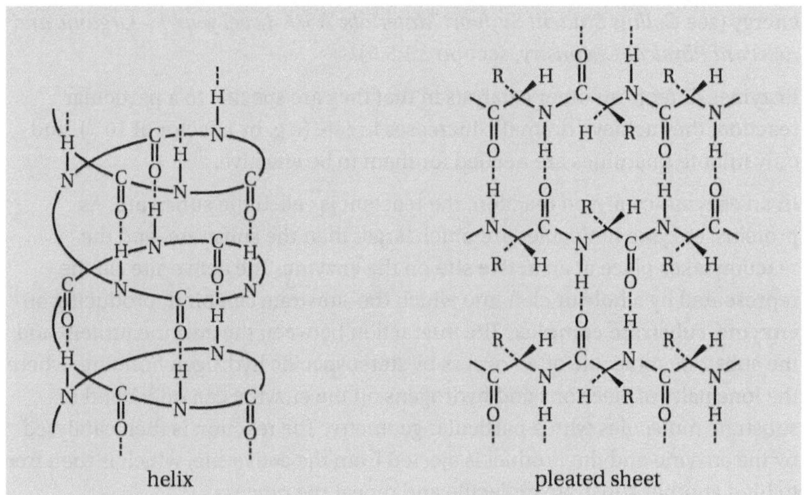

Fig 41
Types of secondary structure of proteins

Disulfide bridges in proteins

Different regions of secondary structure may be present in any one protein. For example, it may have an α-helix in one region and β-pleated sheets in another. Whatever the secondary structure(s), the overall three-dimensional shape of the whole protein is called its **tertiary structure.**

The various electrostatic or covalent links between the R side-chains of amino acid residues in the protein determine how the different regions of secondary structure interact with one another and fold to build up the tertiary structure. It is the tertiary structure that determines the characteristic biochemical activity of the protein.

If a protein contains side-chain —SH groups, which are close enough to one another, these groups can react to form strong S—S covalent bonds called **disulfide bridges.** The disulfide bridges play an important role in fixing the tertiary structure of the protein. Since disulfide bridges can be destroyed by heating, reduction or reaction with a base, the tertiary structure of the protein, and hence its biological activity, can be lost and the protein is then said to be denatured (Fig 42).

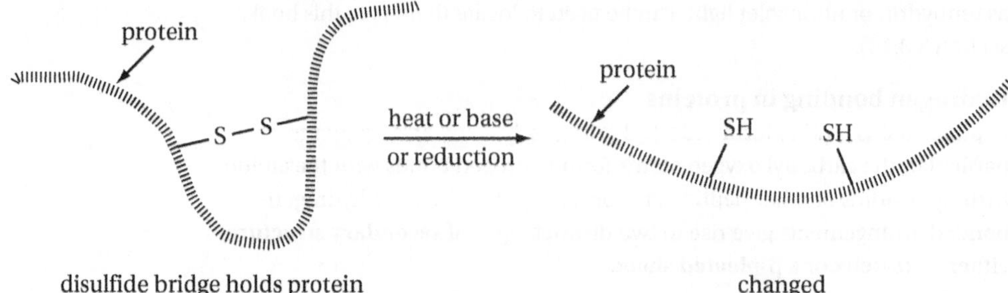

Fig 42 The effect of destroying a disulfide bridge in a protein

3.3.13.3 Enzymes

Enzymes are proteins that act as catalysts for biochemical reactions. Like other catalysts, they increase the rate of reaction by lowering the activation energy (see *Collins Student Support Materials: AS/A-Level year 1 - Organic and Relevant Physical Chemistry*, section 3.1.5.5).

Enzymes differ from other catalysts in that they are specific to a particular reaction: they achieve dramatic increases in rate (e.g. by a factor of 10^{10}), and only minute quantities are needed for them to be effective.

In an enzyme-catalysed reaction, the reactant is called the **substrate**. As proteins, enzyme molecules are much larger than the substrate, and the reaction takes place at an **active site** on the enzyme. The active site can be represented by a hole or cleft into which the substrate can bind, producing an **enzyme–substrate complex**. The interaction between the enzyme protein and the substrate molecule often occurs by stereospecific hydrogen bonding, where the lone pairs of electrons and hydrogens on the enzyme can only bond to substrate molecules with a particular geometry. The reaction is then catalysed by the enzyme and the product is ejected from the active site, which is then free to bind another substrate molecule and repeat the process.

Since enzymes are proteins with three-dimensional tertiary structure, active sites have a unique shape and only molecules with a complementary shape can fit in. This mode of action, known as the **lock and key hypothesis**, is one reason why enzymes are so specific in their catalytic action (see Fig 43).

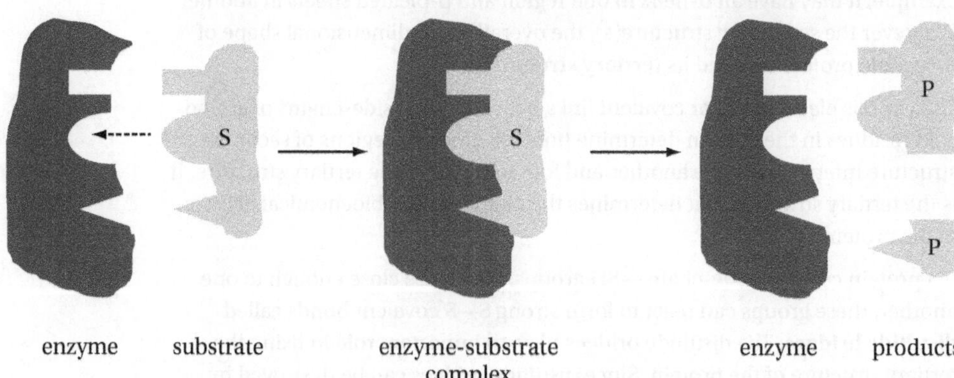

Fig 43 Enzyme–substrate formation and break up

Like the amino acids from which they are made up, enzymes are chiral molecules (see this book, section 3.3.7). This makes enzymes **stereospecific** in their action, in that only one optical isomer of a chiral substrate can bind successfully to active site (see Fig 44).

The pain killer ibuprofen is a good example of stereospecificity, as only one enantiomer is effective. An important lesson about stereospecificity arose out of the notorious thalidomide tragedy of the 1960s. This medicine, used for morning sickness in pregnant women, later gave rise to severe birth defects in babies. The prescribed version of the drug was a racemic mixture of optical isomers. One optical isomer was effective as an anti-nausea treatment, but the other caused the birth defects by interfering with DNA. Thalidomide can inhibit new blood vessels forming in and around tumours and today, under tightly controlled conditions, it is used in the treatment of leprosy and bone cancer.

Separate evaluation of the effectiveness and also of the side effects of both optical isomers of potential new drugs is therefore always vitally important.

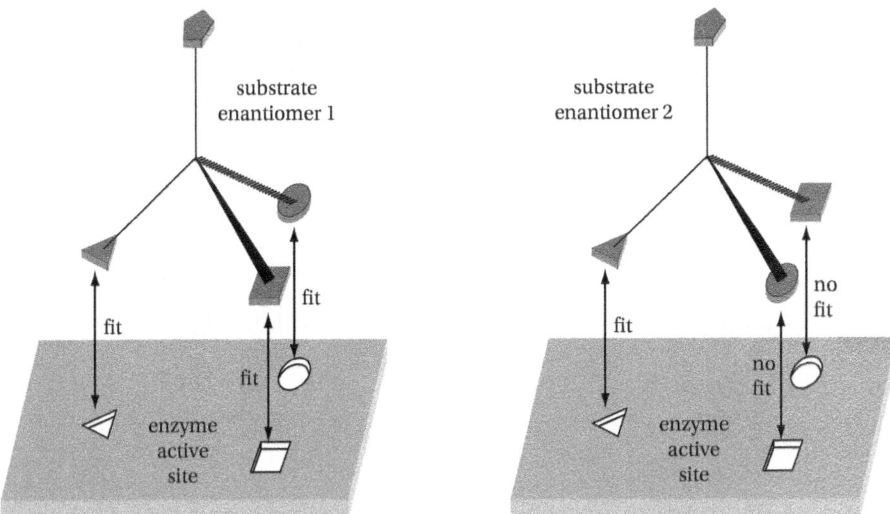

Fig 44
Stereospecific action of an enzyme active site

A molecule that reduces the activity of an enzyme by binding to the active site is called an **enzyme inhibitor**. Many drugs used to treat disease are enzyme inhibitors. The drug molecules bind more strongly to the active site than the natural substrate, so they block the active site and prevent the natural substrate from reacting.

As the three-dimensional shape of an enzyme is the key to its catalytic action and to any attempt to prevent an enzyme functioning, it follows that detailed three-dimensional computer modelling of active sites and the shapes of potential drugs is an important tool in the design of new drugs.

3.3.13.4 DNA

The double helix structure of DNA is familiar to many. Each strand of DNA is a polymer of **nucleotides**. Each nucleotide is made up of a phosphate ion linked by a covalent bond to a **2-deoxyribose sugar** which in turn is covalently bonded to a **base** (one of adenine or guanine (purines) or cytosine or thymine (pyrimidines)). These structures are shown in Figs 45, 46 and 47.

Notes

Fig 46 shows that the only difference between 2-deoxyribose and ribose is the absence of a 2-hydroxy group on 2-deoxyribose. Note that although 2-deoxyribose is regarded as a carbohydrate, it does not fit the general formula of $C_nH_{2n}O_n$. It is classified as a pentose since it has a five-membered carbon ring, and as a single sugar it is a monosaccharide.

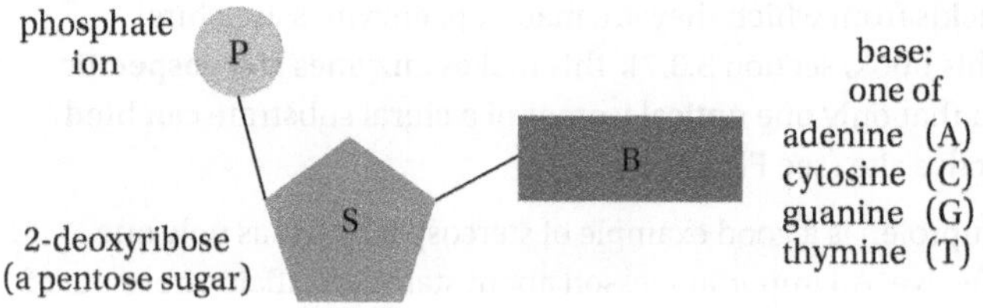

phosphate ion P

2-deoxyribose (a pentose sugar) S

B

base: one of adenine (A) cytosine (C) guanine (G) thymine (T)

Detailed structures are provided in the AQA Data Booklet

Fig 45
Schematic representation of a nucleotide. P = phosphate, S = deoxyribose sugar, B = base

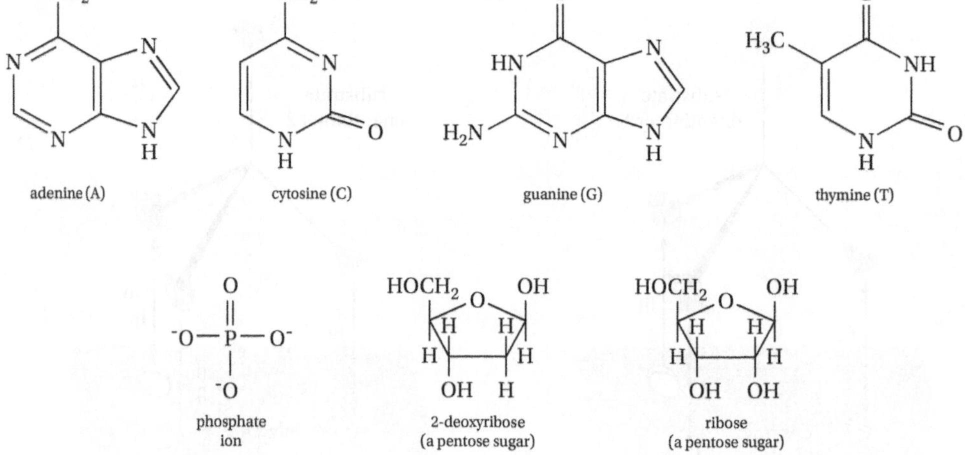

adenine (A)

cytosine (C)

guanine (G)

thymine (T)

phosphate ion

2-deoxyribose (a pentose sugar)

ribose (a pentose sugar)

Fig 46
Detailed structures of the four DNA bases, the phosphate ion and 2-deoxyribose. These are provided in the AQA data booklet. Ribose is shown for comparison with 2-deoxyribose

Notes

Note that in Fig 47, adenine is drawn differently from the AQA data booklet. The structure of all the bases can be written either way round in this way.

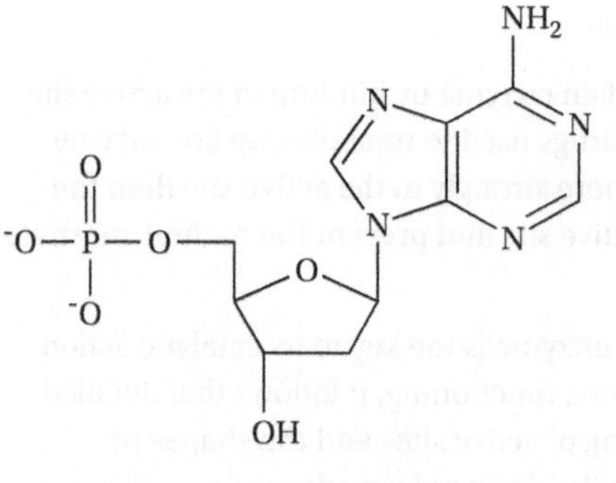

Fig 47
Detailed structure of a single nucleotide

Each DNA strand is a polynucleotide, made up of a **sugar**-phosphate-sugar-phosphate polymer, with a base branching off each sugar molecule. The second strand of the double helix is another sugar-phosphate-sugar-phosphate polymer running in the opposite direction: the two strands are said to be antiparallel. Hydrogen bonding between pairs of bases on the adjacent strands holds the two strands together to form the double helix (Fig 48).

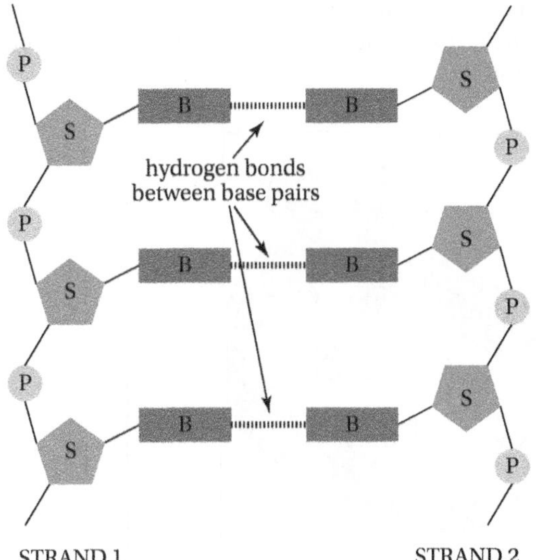

Fig 48
A schematic representation of the DNA double helix

In DNA, adenine pairs with thymine via two hydrogen bonds, whilst cytosine pairs with guanine via three hydrogen bonds (Fig 49).

Each strand complements the other. For example, the base sequence ATCGATTGAGCTCTAGCG links to the other strand with the base sequence TAGCTAACTCGAGATCGC.

When cells containing DNA divide, the DNA double helix (with strand A and **complementary strand** B) unwinds. As the DNA strands unwind, each strand is copied producing two new identical double helices, one containing the original strand A and the other its complementary strand B. This copying of a DNA double helix is called **replication.** Replication of DNA occurs in the development of human embryos and also in the growth of cancer cells.

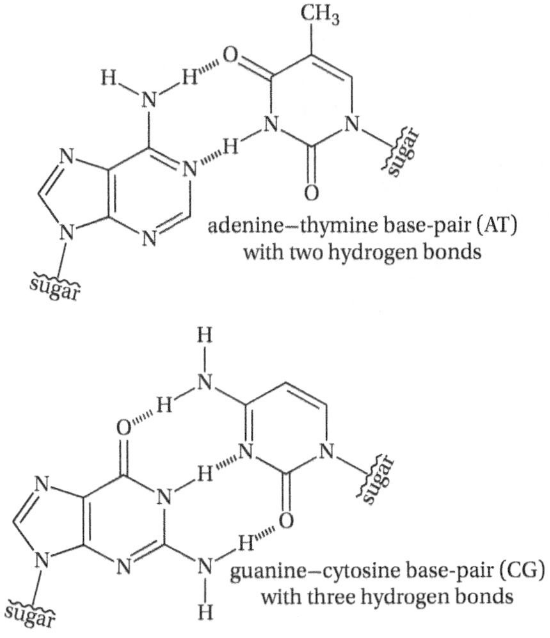

adenine–thymine base-pair (AT) with two hydrogen bonds

guanine–cytosine base-pair (CG) with three hydrogen bonds

Fig 49
Hydrogen bonding in DNA base pairs

The three-dimensional structure of DNA, showing the sugar–phosphate backbone, the helical structure and the hydrogen bonding between the bases is shown in Fig 50.

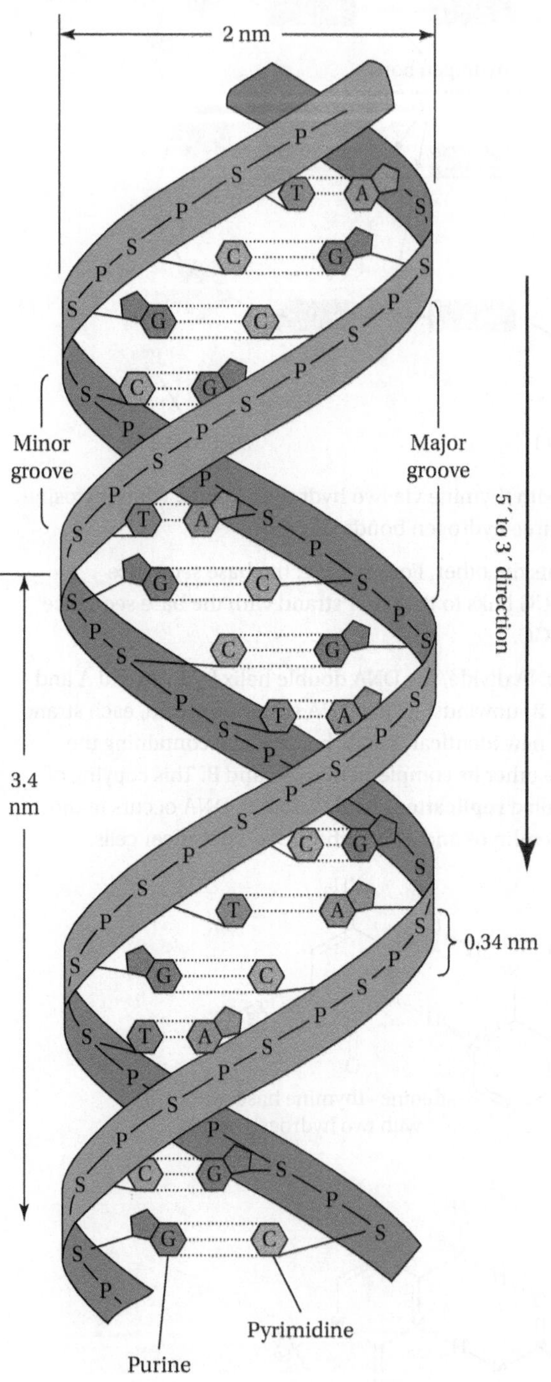

Fig 50
The 3D structure of DNA

3.3.13.5 Action of anticancer drugs

In *Collins Student Support Materials: A-Level year 2 – Inorganic and Relevant Physical Chemistry*, section 3.2.5.4, the *cis–trans* geometrical isomers of the square planar complex diamminedichloroplatinum(II), known as cisplatin and transplatin, were discussed (see Fig 51).

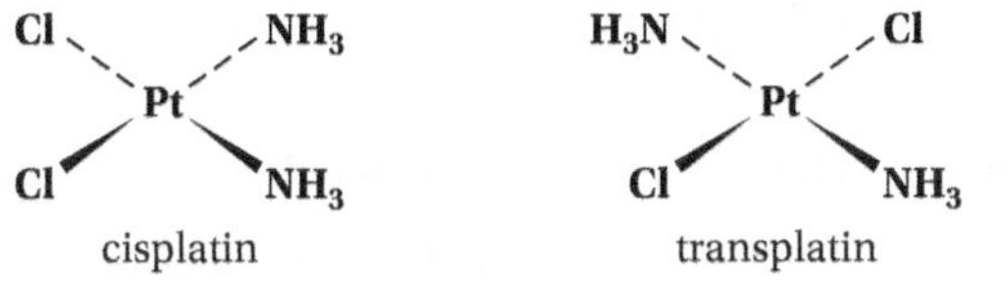

cisplatin transplatin

Fig 51
Platin isomers

Cisplatin, whose structure is on the left in Fig 51, has had remarkable success in chemotherapy for the treatment of certain types of cancer. Cisplatin works by undergoing a ligand exchange reaction with one of the nitrogen atoms of a guanine base on DNA.

The drug is usually injected in the form of $(NH_3)_2PtCl_2$. This neutral complex, after passing through the cell membrane of a cancer cell, reacts with water to form the complex ion $[(NH_3)_2PtCl(H_2O)]^+$.

Even though the guanine is still bonded to the 2-deoxyribose in a DNA strand (see this book, section 3.3.13.4), its nitrogen 7 forms a coordinate bond to the platinum, replacing water (see Fig 52). Another guanine on the same strand of DNA then bonds to the platinum by replacing the remaining chloride ligand (see Fig 53). As a result, the strand becomes kinked and will not replicate.

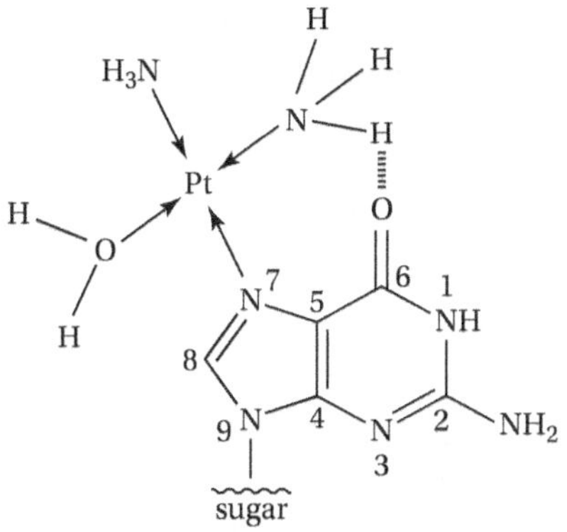

The Cl^- ion of the complex ion $[(NH_3)_2 PtCl(H_2O)]^+$. formed from cisplatin, is replaced by a guanine base on a DNA strand.

Fig 52
Pt complex binds to a single guanine base on DNA

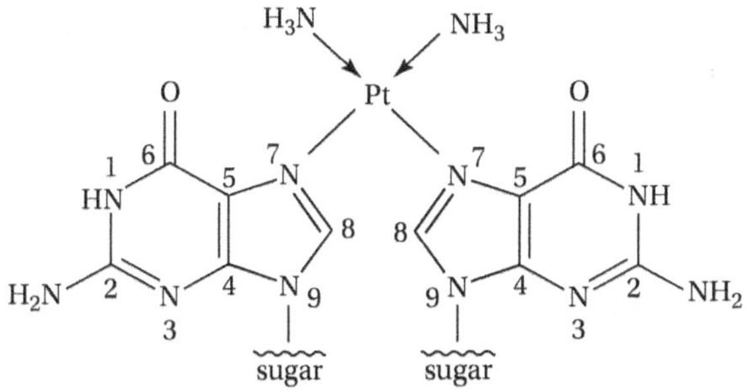

The Pt complex binds to the nitrogen-7 atoms of two guanine bases on the same DNA strand, causing the strand to kink and not replicate.

Fig 53
Pt complex binds to two guanine bases in DNA

3.3.14 Organic synthesis

The synthesis of an organic compound from simple starting materials can involve several steps.

In designing the most efficient synthetic route, chemists must take into account several factors:

- avoiding the use of hazardous starting materials
- limiting the use of solvents which might have environmental impact
- using the minimum number of steps
- using steps with a high **atom economy**.

To achieve these combined objectives, chemists rely on a sound knowledge of functional group interconversions. The Appendix contains a summary of the various organic reactions involved in *Collins Student Support Materials: AS/A-Level year 1 - Organic and Relevant Physical Chemistry* and in this book.

Given a target molecule, a suitable route can often be devised in a stepwise manner by *working backwards from the product,* as well as forwards.

Example

Show how CH_3CH_2Br can be converted into $CH_3COOCH_2CH_3$.

The product, an ester, can be made from ethanoic acid and ethanol.
Hydrolysis of CH_3CH_2Br will give ethanol which can, in turn, be oxidised to ethanoic acid. A likely reaction sequence is:

$$CH_3CH_2Br \xrightarrow{KOH(aq)} CH_3CH_2OH \xrightarrow{[O]} CH_3COOH$$

$$CH_3CH_2OH + CH_3COOH \xrightarrow{conc.\ H_2SO_4} CH_3COOCH_2CH_3$$

Example

Show how $CH_3CH{=}CH_2$ can be converted into $(CH_3)_2CHCH_2NH_2$.

The CH_2NH_2 section of the product suggests the reduction of a CN group as the final step.
The CN group is likely to have been introduced by a nucleophilic substitution, i.e.

$RBr \rightarrow RCN$

Addition of HBr to propene must be the first step.
The reaction sequence is therefore:

$CH_3CH{=}CH_2 + HBr \rightarrow$

$$CH_3CH(Br)CH_3 \xrightarrow{KCN} CH_3CH(CN)CH_3 \xrightarrow{Ni/H_2} CH_3CH(CH_2NH_2)CH_3$$

Example

Outline a reaction sequence for the synthesis of $CH_3COCH_2NHCOCH_3$, starting from ethanal, using appropriate reagents.

The product contains a ketone group and a substituted amide.
Working backwards, the following conversions should come to mind:

$$CH_3COCH_2NH_2 \xrightarrow[(CH_3CO)_2O]{CH_3COCl \text{ or}} CH_3COCH_2NHCOCH_3$$

(acylation of an amine)

$$CH_3CH(OH)CH_2NH_2 \xrightarrow{[O]} CH_3COCH_2NH_2$$

(oxidation of a secondary alcohol)

$$CH_3CH(OH)CN \xrightarrow[\text{or } LiAlH_4]{H_2/Ni} CH_3CH(OH)CH_2NH_2$$

(reduction of a nitrile)

$$CH_3CHO \xrightarrow{KCN/\text{dilute acid}} CH_3CH(OH)CN$$

(nucleophilic addition of HCN)

The reaction sequence is therefore:

$$CH_3CHO + KCN/\text{dilute acid} \rightarrow CH_3CH(OH)CN \rightarrow CH_3CH(OH)CH_2NH_2 \rightarrow CH_3COCH_2NH_2 \rightarrow CH_3COCH_2NHCOCH_3$$

3.3.15 Nuclear magnetic resonance spectroscopy

Many atomic nuclei behave as if they were spinning and are said to have **nuclear spin**. Such nuclei have a **magnetic moment,** which means that they behave like tiny bar magnets. Important examples include ^{1}H (called a proton in an NMR context), and ^{13}C.

When an external magnetic field is applied, the nuclei, which have spin, line up in the same direction (with the field) or in the opposite direction (against the field). The nuclei aligned with the field are lower in energy than those against the field. A signal is recorded when a nucleus aligned with the magnetic field absorbs low-energy radiation in the radio-frequency range and the nucleus enters the higher energy state, which causes **resonance**.

Essential Notes

Nuclei possessing even numbers of both protons and neutrons, such as ^{12}C or ^{16}O, lack magnetic properties and do not give rise to NMR signals.

Notes

Note that the word 'proton' as used in NMR spectroscopy is not an H^+ ion, but refers to a magnetically responsive hydrogen atom – whose nucleus is a proton.

Notes

Examination questions involving NMR will not require a knowledge of NMR theory. Candidates will only need to apply their understanding of NMR spectra to problem solving.

^1H NMR spectroscopy

Protons in organic molecules

In organic molecules, protons are surrounded by electrons which partly shield them from the applied magnetic field. The amount of shielding depends on the electron density surrounding the nucleus and varies for different protons within a compound. Factors which influence the electron density include bond polarity and the presence of electron-donating or electron-withdrawing groups:

- The nucleus is **deshielded** when the electron density is reduced.
- The nucleus is **shielded** when the electron density is increased.

Because each chemically distinct hydrogen atom (proton) has a unique electronic environment, it gives rise to a characteristic resonance. *Chemically equivalent protons* are all in the same environment and therefore absorb at the same frequency. For hydrogen atoms, the differences in frequency are tiny, being recorded as only a few **parts per million** (ppm).

Chemical shift

The movements caused by shielding (moving the recorded value *upfield*) and by deshielding (moving the recorded value *downfield*) are quantified by the **chemical shift**, δ. Chemical shifts are measured, in ppm, relative to an internal standard, tetramethylsilane (TMS), $(CH_3)_4Si$, because:

- TMS gives a signal that resonates upfield from almost all other organic hydrogen resonances, because the 12 equivalent hydrogens are highly shielded.
- TMS gives a single, intense peak since there are 12 equivalent protons.
- TMS is non-toxic and inert.
- TMS has a low boiling point (26.5 °C) and can easily be removed from a sample.

By definition, the δ value of $(CH_3)_4Si$ is zero. Almost all proton NMR absorptions occur 0–10 δ downfield from $(CH_3)_4Si$ (Fig 54). The δ values vary according to the *structural* environment, so that organic functional groups have characteristic chemical shifts.

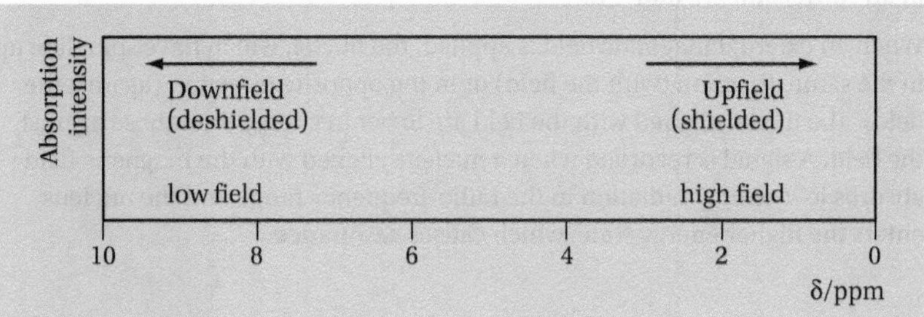

Fig 54
The δ scale of chemical shifts

Proton NMR spectra are recorded in solution. The sample to be examined (a few milligrams) is dissolved in a proton-free solvent to avoid unwanted absorptions. Typical solvents include CCl_4 and deuterated compounds, such as $CDCl_3$ and C_6D_6.

Features of proton NMR spectra

There are four important aspects of ^1H NMR spectra which provide information about chemical structure:

- The *number of absorptions* indicates how many kinds of non-equivalent protons are present.
- The *intensities of the absorptions* reveal how many protons are associated with each resonance peak.
- The *positions of the absorptions* give clues as to the environment of each kind of proton.
- The *splitting of an absorption* into several peaks, which is caused by **spin–spin coupling**, provides information about neighbouring protons.

High-resolution proton NMR spectroscopy has become a very powerful tool for elucidating organic chemical structures.

Essential Notes

High-resolution spectroscopy is able to reveal details of the structure of a peak that are obscured in low-resolution spectroscopy.

Number of absorptions

The different environments of the protons in an organic molecule are revealed by an examination of its low-resolution ^1H NMR spectrum. In the case of ethanol, for example, three distinct signals are found (Fig 55).

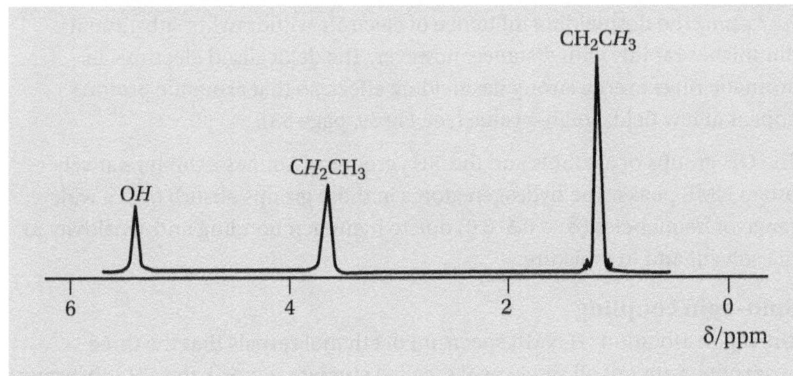

Fig 55
Low-resolution ^1H NMR spectrum of ethanol, CH_3CH_2OH

The three kinds of non-equivalent proton are seen at different positions and the peaks are of different intensities. For each peak, the area is proportional to the number of hydrogen atoms giving rise to the signal. Thus, for ethanol, the areas under the peaks are in the ratio of 1:2:3, in accordance with HO:CH_2:CH_3. The spectrometer is able to measure the **relative intensities** of the various peaks electronically and provide an **integration trace**, which calculates the area under each peak. Integrated peak areas are sometimes presented on the recorded spectrum in a stepwise manner, so that the height of each step is proportional to the number of hydrogen atoms associated with each signal (see Fig 57, page 51).

Position of absorption

As pointed out earlier, the effects of shielding and deshielding are expressed by chemical shift values on the δ scale. For ethanol (Fig 55), the hydroxyl hydrogen atom is deshielded because of the effect of the electron-withdrawing electronegative oxygen atom. However, the CH_2 and CH_3 groups are progressively further from the oxygen atom and the electron density around the hydrogen atoms increases. The CH_3 hydrogen atoms are the most highly shielded and absorb at the higher field.

In general, the positions of absorption (as measured by chemical shifts) can be related to the electronegativities of adjacent atoms and the electron-withdrawing or electron-donating effects of functional groups (Table 5).

Table 5
Some typical proton chemical shifts

Type of proton	*δ/ppm*
RCH_3	0.8–1.0
R_2CH_2	1.2–1.4
R_3CH	1.4–1.6
$RCOCH_3$	2.1–2.6
RCH_2OR	3.3–3.9
RCH_2Br	3.4–3.6
RCH_2Cl	3.6–3.8
$R_2C{=}CH_2$	4.6–5.0
ArH	6.0–8.5
RCHO	9.5–9.9

Notes

Note that the δ values in Table 5 are only approximate because they are affected by neighbouring substituents. Appropriate data will be provided, as necessary, for use in examination questions.

Multiple substitution has a cumulative effect. Thus, for example, the δ values for CH_3Cl, CH_2Cl_2 and $CHCl_3$ are, respectively, 3.05, 5.30 and 7.27 ppm. The deshielding influence of electron-withdrawing substituents diminishes rapidly with distance, however. The delocalised electrons in aromatic rings exert a strong deshielding effect, so that aromatic protons appear at low field – high δ value (see Fig 59, page 53).

The OH groups of alcohols and the NH_2 groups of amines exhibit relatively broad NMR peaks. The hydrogen atoms in these groups absorb over a wide range of frequencies (δ = 0.5–6.0) due to hydrogen bonding and sensitivity to the solvent and to moisture.

Spin–spin coupling

Notes

Note that equivalent hydrogens on adjacent atoms do not display any coupling effects.

The high-resolution ^1H NMR spectrum of ethanol reveals that the three absorptions are not all single peaks, called **singlets** (Fig 56): the CH_3 absorption signal is split into three peaks, a **triplet**, whereas the CH_2 signal appears as four peaks, a **quartet**. This situation arises because non-equivalent hydrogens on adjacent atoms interact – *couple* – with one another.

In general, **splitting** of single absorption peaks into more complex patterns of so-called **multiplets** is due to coupling between neighbouring nuclear spins. Thus, the spin of one proton can couple with the spins of adjacent protons. Splitting is observed only between nuclei with *different* chemical shifts.

The splitting of an absorption signal is described by the ***n* + 1 rule**:

Definition

Signals for protons adjacent to n equivalent neighbours split into n + 1 peaks.

For a molecule such as $ClCH_2CHCl_2$, the two NMR signals (from the CH_2 and CH protons) are split into a **doublet** and a **triplet**, respectively (Fig 57). The

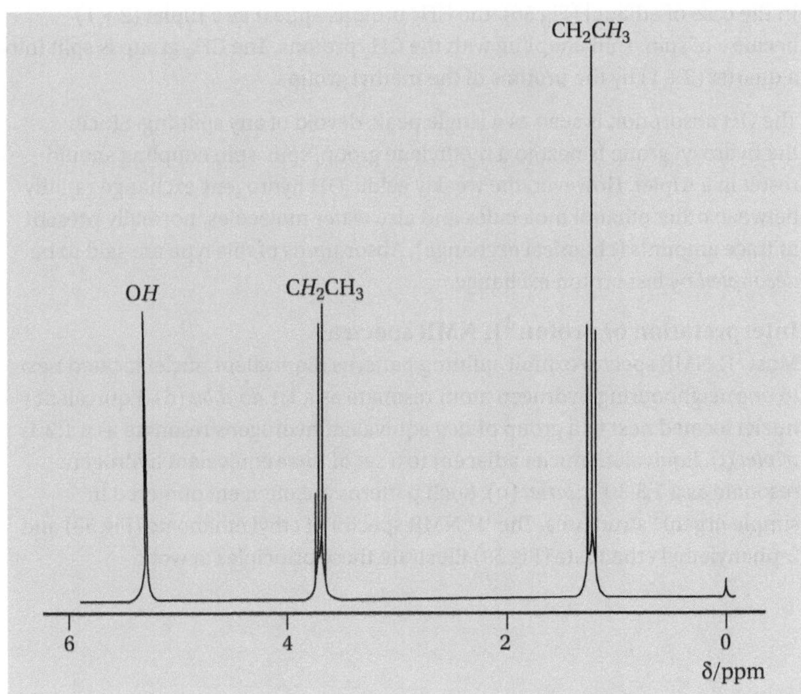

Fig 56
High-resolution ^1H NMR spectrum of ethanol, CH_3CH_2OH

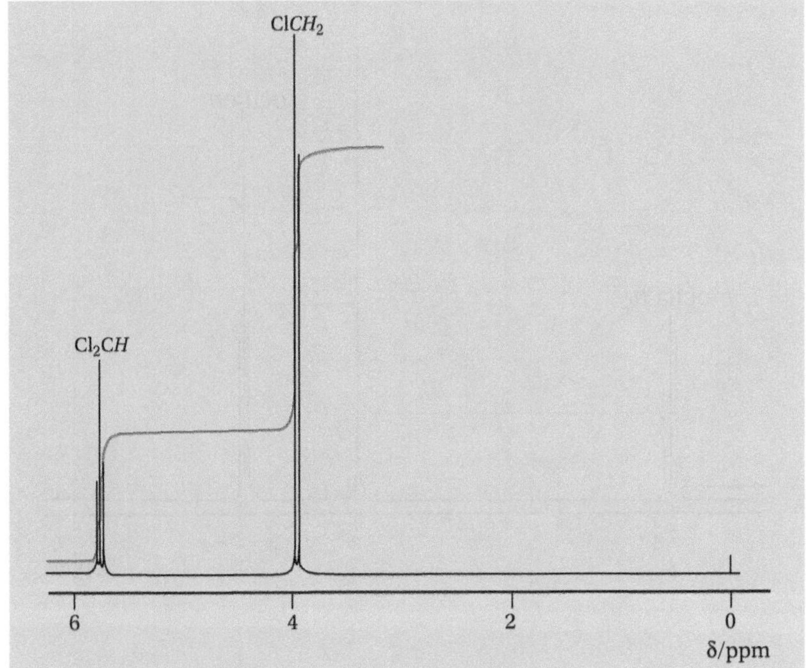

Fig 57
^1H NMR spectrum of 1,1,2-trichloroethane, $ClCH_2CHCl_2$

peak for the two equivalent CH_2 protons is split into a doublet (1 + 1) by the single adjacent proton. The peak for the CH proton, however, is split into a triplet (2 + 1) by the two CH_2 protons.

Notes

The multiplicity and relative intensities of the $n + 1$ components are given by the coefficients of the terms in the expansion of $(1 + x)^n$. These values can be obtained from Pascal's triangle

1

1 1

1 2 1

1 3 3 1

1 4 6 4 1

1 5 10 10 5 1

1 6 15 20 15 6 1

where each inner number is the sum of the two numbers closest to it in the row above.

In the case of ethanol (Fig 56), the CH_3 protons appear as a triplet (2 + 1) because of spin–spin coupling with the CH_2 protons. The CH_2 group is split into a quartet (3 + 1) by the protons of the methyl group.

The OH absorption is seen as a single peak, devoid of any splitting. Since the hydroxyl group is next to a methylene group, spin–spin coupling should result in a triplet. However, the weakly acidic OH hydrogens exchange rapidly between other ethanol molecules and also water molecules, normally present in trace amounts (chemical exchange). Absorptions of this type are said to be *decoupled* by fast proton exchange.

Notes

Questions about splitting in AQA examinations will be restricted to a consideration of protons that have 1, 2 and 3 neighbouring hydrogen atoms only, giving rise to doublets, triplets and quartets, respectively.

Interpretation of proton ^1H NMR spectra

Most ^1H NMR spectra exhibit splitting patterns. Equivalent nuclei located next to *one* neighbouring hydrogen atom resonate as a 1:1 *doublet* (d). Equivalent nuclei located next to a group of *two* equivalent hydrogens resonate as a 1:2:1 *triplet* (t). Equivalent nuclei adjacent to a set of *three* equivalent hydrogens resonate as a 1:3:3:1 *quartet* (q). Such patterns are often encountered in simple organic structures. The ^1H NMR spectra of ethyl ethanoate (Fig 58) and 2-phenylethyl ethanoate (Fig 59) illustrate these principles at work.

Fig 58
^1H NMR spectrum of ethyl ethanoate, $CH_3COOCH_2CH_3$

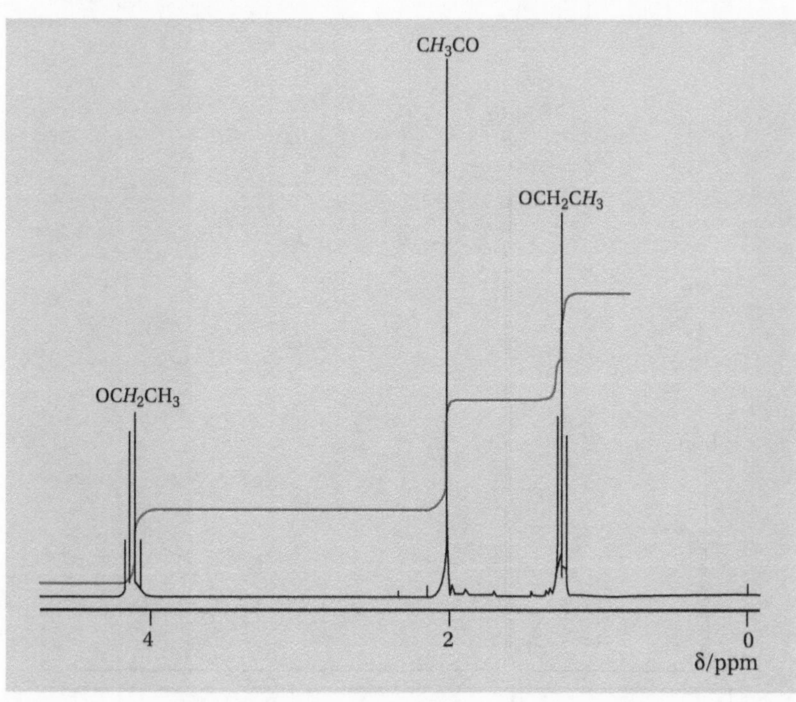

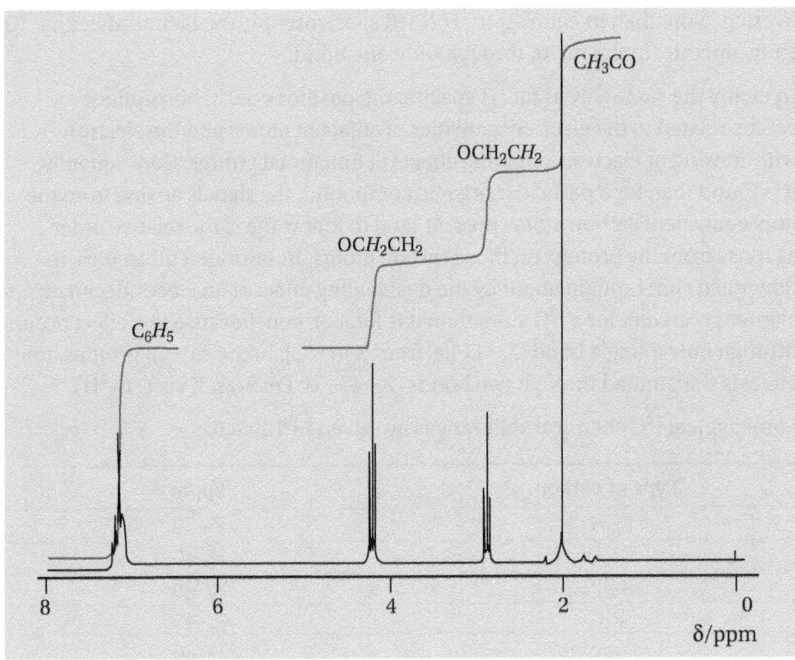

Fig 59
^{1}H NMR spectrum of 2-phenylethyl ethanoate, $CH_3COOCH_2CH_2C_6H_5$

Notes
Examination questions involving the spin–spin coupling of aromatic protons will not be set.

^{13}C NMR spectroscopy

^{13}C NMR spectroscopy provides direct information about the carbon skeleton of a molecule, and it is possible to determine:

- The *number* of non-equivalent carbon atoms in the structure (i.e. the *number* of carbon atoms in different chemical environments).
- The *different types* of carbon atom present in the compound (typically saturated, unsaturated, aromatic and carbonyl carbon atoms).

The interpretation of ^{13}C NMR spectra is, in general, easier than in the case of ^{1}H NMR spectra. However, both techniques are often used in combination in structure elucidation, usually together with infra-red and mass spectral data.

The ^{13}C isotope is magnetically active, in the same way as ^{1}H, but its natural abundance is only 1.1%. The resonances of ^{13}C nuclei are therefore more difficult to observe and are much weaker than proton resonances. Consequently, a ^{13}C NMR spectrum has to be built up from a collection of molecules, since an individual molecule is unlikely to contain more than one ^{13}C nucleus and is therefore unable to provide more than a single ^{13}C resonance. Thus, a greater number of individual scans of the spectrum must be accumulated than is the case for a proton spectrum.

As for ^{1}H spectra, the **chemical shift** (δ) is an important parameter. By definition, the δ value of the internal standard tetramethylsilane, $(CH_3)_4Si$, is zero, as it is for proton NMR. Here, however, it is the methyl group *carbon* atoms that are used for reference, as opposed to the methyl group *hydrogen* atoms. Each non-equivalent carbon atom in an organic molecule gives rise to a signal with a quite different ^{13}C chemical shift; the typical range of observed chemical shifts (0 to 220 ppm) in ^{13}C NMR spectroscopy is much larger than that for protons (0 to 12 ppm). Because of this wide range of values, and in contrast to many proton NMR spectra, the different ^{13}C peaks are less likely to overlap. Note that, in contrast to ^1H NMR spectroscopy, the factors affecting ^{13}C shifts operate, in the main, through only one bond.

In nearly the same way as for ^1H spectra, the positions of ^{13}C absorptions can be related to the electronegativities of adjacent atoms and the electron-withdrawing or electron-donating effects of functional groups. Consequently, it is found that, for a particular organic compound, the signals arising from the non-equivalent carbon atoms present tend to follow the same relative order as those from the protons on those carbon atoms. In saturated molecules, the downfield shift brought about by the deshielding effect of an electronegative element is greater for a ^{13}C atom than it is for a proton, because this effect occurs through only a single bond (X—C, i.e. from X to ^{13}C); in the case of protons, the effect is transmitted through two bonds (X—C—H, i.e. from X via C to ^1H).

Some typical ^{13}C chemical shift ranges are given in Table 6.

Table 6
Some typical ^{13}C chemical shift ranges

Type of carbon	*δ/ppm*
R$\underline{C}H_3$	5–30
$R_2\underline{C}H_2$	15–40
$R_3\underline{C}H$	20–40
$R_4\underline{C}$	25–40
$\underline{C}$N (amine)	25–60
$\underline{C}$C=O	20–50
$\underline{C}$Br	10–60
$\underline{C}$Cl	25–70
$\underline{C}$OR (alcohol, ether)	50–75
O=CO$\underline{C}$ (ester)	50–80
$\underline{C}$≡N (nitrile)	110–125
$\underline{C}$=$\underline{C}$ (alkene)	100–150
$\underline{C}$=$\underline{C}$ (aromatic)	110–160
$\underline{C}$=O (ester, amide, acid)	160–185
$\underline{C}$=O (aldehyde, ketone)	190–220

Notes
The AQA Data Booklet contains a simplified version of this list for use in examinations.

Due to the low natural abundance of ^{13}C, spin-spin coupling between two adjacent ^{13}C atoms in the same molecule is extremely unlikely and can be ignored; the chance of finding more than one ^{13}C atom in a molecule is almost nil.

Although spin-spin coupling between ^{13}C and ^1H does occur, most ^{13}C NMR spectra are obtained as **proton-decoupled spectra**, in which only singlets are observed for each of the non-equivalent carbon atoms present. This instrumental technique greatly simplifies the spectrum and avoids overlapping multiplets but, as a consequence, all the information on the number of attached hydrogen atoms is lost. At the same time, the intensities of some of the carbon resonances increase beyond those observed in the corresponding proton-coupled spectrum; such enhancement increases, but not always linearly, with the number of hydrogen atoms attached. This enhancement is called the nuclear Overhauser effect and is named after its discoverer.

Notes
You are **not** required to know about proton-coupled spectra. Examination questions will only refer to proton-decoupled spectra, with no splitting of the ^{13}C peaks.

Integral information derived from ^{13}C NMR spectra is not directly proportional to the number of atoms giving rise to the signal. However, a singlet peak derived from two (equivalent) carbon atoms is larger than one derived from a single carbon atom. It is often, but not always, found that a CH_3 peak has a greater intensity than a CH_2 peak, which in turn is more intense than a CH peak; quaternary carbon atoms are normally the weakest peaks in the spectrum.

The proton-decoupled ^{13}C NMR spectrum of methyl ethanoate is shown in Fig 60; the assignments given are consistent with the values provided in the chemical shift table (Table 6). Thus, the methyl carbon atom (*c*), adjacent to the ester carbonyl group, is found at high field (δ = 20.6 ppm), whereas the methyl carbon atom (*b*), next to the electronegative oxygen atom, is further downfield (δ = 51.6 ppm); the ester carbonyl group (*a*) appears as a singlet at low field (δ = 171.5 ppm).

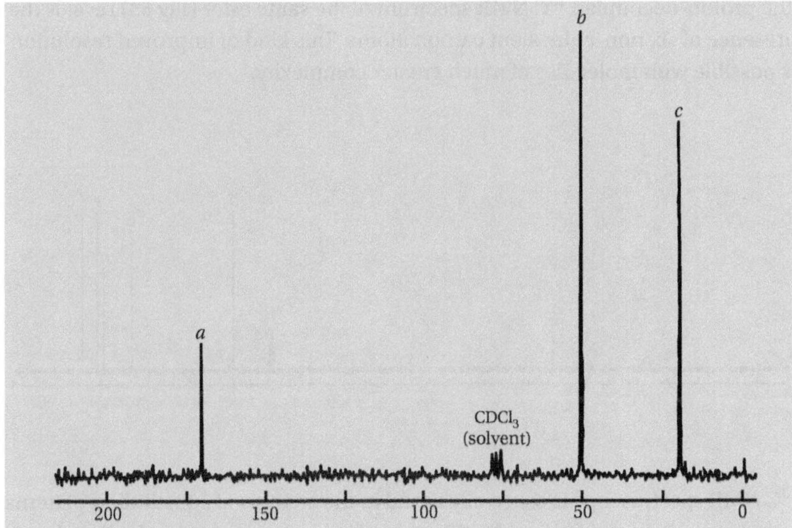

Fig 60
Proton-decoupled ^{13}C NMR spectrum of methyl ethanoate

CH_3COOCH_3
c *a* *b*

a = 171.5 ppm
b = 51.6 ppm
c = 20.6 ppm

Comparisons between the ^1H and the ^{13}C NMR spectra of the same compound can be useful and instructive. Ethyl ethanoate provides a simple example. The ^1H NMR spectrum of this ester appears in Fig 58, revealing three non-equivalent groups of hydrogen atoms. The corresponding proton-decoupled ^{13}C NMR spectrum (Fig 61) shows the presence of four non-equivalent carbon atoms in the structure; the signal at approximately 77 ppm is that of the solvent ($CDCl_3$).

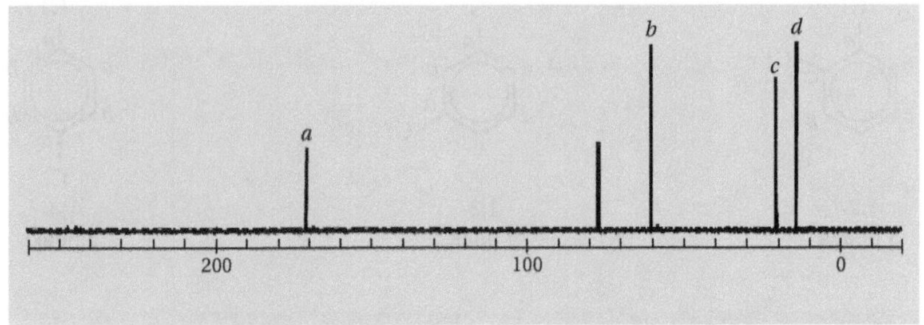

Fig 61
Proton-decoupled ^{13}C NMR spectrum of ethyl ethanoate

$CH_3COOCH_2CH_3$
c *a* *b* *d*

a = 171.1 ppm
b = 60.4 ppm
c = 21.0 ppm
d = 14.3 ppm

The relative simplicity of ^{13}C NMR spectra is useful in the identification of increasingly large alkyl groups, where complex multiplets are observed in the corresponding ^1H NMR spectra. For example, the ^1H NMR spectrum of butyl ethanoate is shown in Fig 62. There are five non-equivalent groups of protons present in this ester, giving rise to a singlet (δ = 1.87 ppm), two triplets (δ = 3.90 ppm and δ = 0.78 ppm) and two unresolved multiplets (which are centred at δ = 1.45 ppm and δ = 1.23 ppm).

Fig 62
^{1}H NMR spectrum of butyl ethanoate

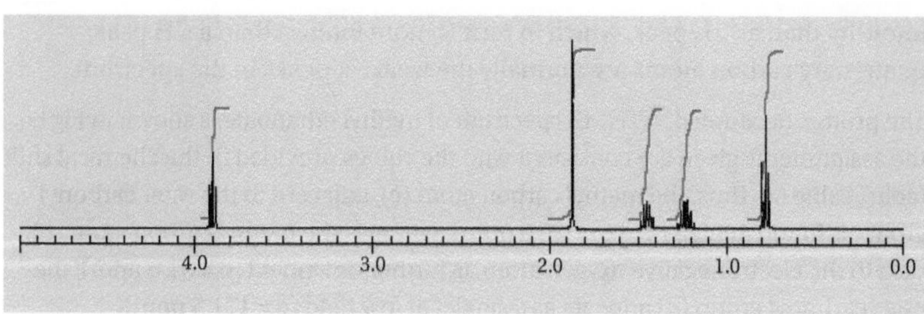

The proton-decoupled ^{13}C NMR spectrum of the same ester (Fig 63) reveals the presence of six non-equivalent carbon atoms. This kind of improved resolution is possible with molecules of much greater complexity.

Fig 63
Proton-decoupled ^{13}C NMR spectrum of butyl ethanoate

$CH_3COOCH_2CH_2CH_2CH_3$
d a b c e f

a = 170.9 ppm
b = 64.1 ppm
c = 30.6 ppm
d = 20.8 ppm
e = 19.0 ppm
f = 13.5 ppm

^{13}C NMR spectroscopy is especially useful in the analysis of substitution patterns in benzene rings and for the identification of isomers. Most polysubstituted benzene rings display six different peaks in the proton-decoupled spectrum, one for each carbon atom; symmetrical substitution can reduce this number. Dichlorobenzene provides a straightforward example. Symmetry considerations reveal that 1,2-dichlorobenzene has three unique carbon atoms (3 peaks), 1,3-dichlorobenzene has four non-equivalent carbon atoms (4 peaks) whereas 1,4-dichlorobenzene has only two such atoms (2 peaks):

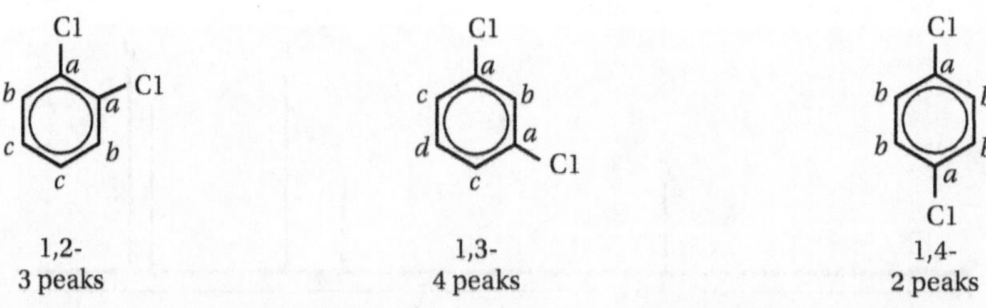

3.3.16 Chromatography

Chromatography is the collective term used for various related laboratory techniques that permit the separation and identification of the chemical components in a mixture. All forms of chromatography involve a fixed **stationary phase** through which passes a **mobile** or **moving phase** containing the mixture to be separated. Separation is achieved because components of the mixture distribute themselves differently between the two phases according to their affinity for each phase.

Definition

***Chromatography** is a technique for separating the components of a mixture on the basis of differences in their affinities for a stationary and for a moving phase.*

It is helpful to make a distinction between **analytical** chromatography and **preparative** chromatography. Analytical chromatography operates with small amounts of material and aims to identify and measure the relative proportions of the various components present in a mixture; such an approach often involves comparisons with known standards for purposes of identification. Preparative chromatography is carried out on a relatively large scale and is a form of purification.

Notes

The term **chromatography** is coined from two Greek words, χρωμα (chroma), meaning colour, and γραφειν (graphein), meaning to write. The word was first used in 1906 by the Russian botanist Mikhail Tsvet, who invented the technique in 1901 when he separated chlorophyll and carotenoid plant pigments down a column of calcium carbonate.

Two main kinds of chromatography can be distinguished:

- **partition** chromatography
- **adsorption** chromatography.

In partition chromatography, the stationary phase is a thin, non-volatile liquid film held on the surface of an inert solid or within the fibres of a supporting matrix; the moving phase is a liquid or a gas. Solute molecules equilibrate, or **partition**, between the two phases. Separation depends on the balance between solute solubility in the moving phase and retention in the stationary phase.

In adsorption chromatography, the stationary phase is a solid, such as alumina (Al_2O_3), and the moving phase is a liquid or a gas. The surface area of the solid phase is maximised by the use of finely-divided particles. Solute molecules become attached to **adsorption** sites on the stationary phase. Strongly adsorbed molecules travel more slowly in the moving phase than those that are only weakly adsorbed.

Types of chromatography included within the two main categories are:

- **paper** chromatography
- **thin-layer** chromatography (TLC)
- **column** chromatography (CC)
- **gas** chromatography (GC).

Paper chromatography

This technique can be regarded as an example of partition chromatography where the stationary phase is essentially a thin layer of water adsorbed on the cellulose fibres of the paper. The moving phase is a solvent or a solvent mixture.

Drops of concentrated sample solution are spotted along a baseline, drawn in pencil near the bottom of a piece of chromatography paper. The dried paper is then dipped in a shallow layer of solvent inside a suitable container and sealed (see Fig 64). The solvent rises up the paper, carrying along the solute components; these travel at different rates according to their different solubilities in the moving and stationary phases, resulting in separation. The

Notes

Paper chromatography is an effective method for revealing the coloured components present in some sweets (typically Smarties) and in ballpoint pen inks or felt-tip markers.

Fig 64
Paper chromatography of three simple mixtures

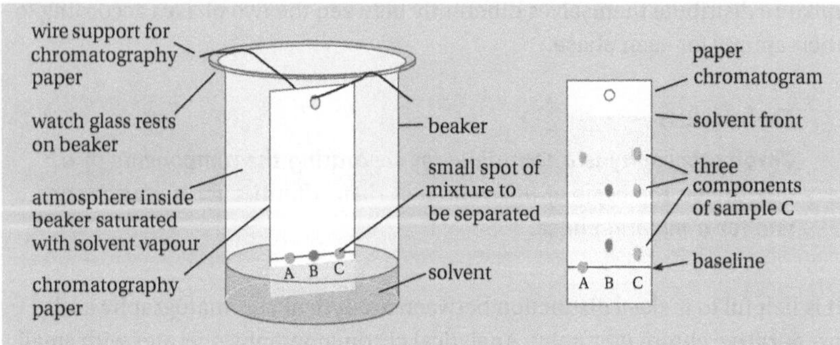

process is stopped when the leading edge of the solvent, the **solvent front**, gets close to the top of the paper. The resulting **chromatogram** is taken out and dried.

Definition

A chromatogram is a pattern of separated substances obtained by chromatography.

The ratio of the distance travelled up the paper by a component relative to that of the solvent is called the **retention factor, R_f**:

$$R_f = \frac{\text{distance travelled by the compound}}{\text{distance travelled by the solvent front}}$$

Under standard conditions, R_f values can be used as a means of identification. The most reliable approach is to compare both known and unknown compounds together on the same chromatogram.

When, as in the majority of cases, the constituents of a mixture are colourless, it is necessary to treat the chromatogram to make the components visible. Heating the chromatogram will sometimes suffice, as will exposure to ultra-violet light.

In some cases, the components of a mixture cannot be **resolved**, i.e. separated completely, by paper chromatography in one direction. The problem can often be overcome by repeating the process using a different solvent, after rotating the paper through 90°, in order to produce a two-dimensional array of components (see Fig 65).

Notes

Two-dimensional paper chromatography is quite powerful and was used by the biochemist Frederick Sanger in 1955 to identify the 17 amino acids generated by the acid hydrolysis of the protein insulin. Amino-acid spots on a chromatogram are made visible as bluish-purple stains by treatment with a solution of the reagent ninhydrin.

Fig 65
Two-dimensional paper chromatography

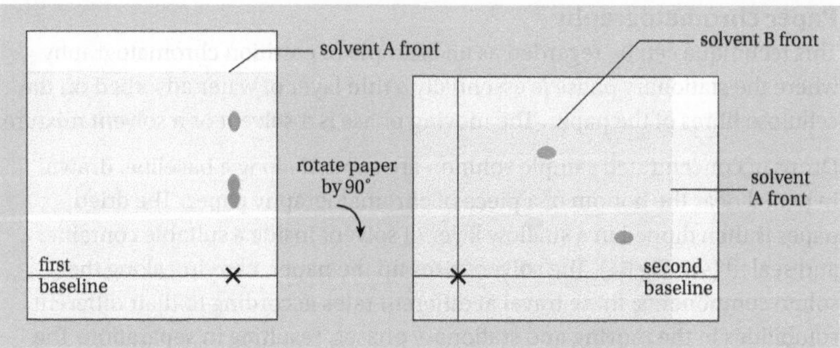

Thin-layer chromatography (TLC)

This laboratory technique is similar to paper chromatography. The stationary adsorbent phase is a thin layer of a polar matrix, such as silica gel or alumina, coated as a paste on to a glass plate or on to a thick aluminium or plastic sheet and then dried. A concentrated solution of the mixture is applied as a spot or a streak near the bottom of the plate. The treated plate is dipped in a shallow layer of solvent (the moving phase) which gradually rises up the plate by capillary action, leading to separation of the components; solutes travel different distances according to how strongly they interact with the adsorbent.

The plate is taken out and dried when the solvent front has almost reached the top. The chromatogram can be used to provide R_f values in the same way as for paper chromatography.

Visualisation of the spots on a chromatogram is often helped by the deliberate addition of a fluorescent inorganic compound to the adsorbent. The coated plate displays an overall pale green fluorescence under ultra-violet exposure; this fluorescence is quenched by the chemicals present in the spots, which renders them visible.

TLC has several advantages over paper chromatography:

- the procedure is quicker
- separations are more efficient
- results are more easily reproduced
- the adsorbent can be varied.

Thin-layer adsorption chromatography is used widely to monitor the course of organic reactions. It is often possible to follow the gradual disappearance of starting material and the appearance of product.

Notes

Separation of species by thin-layer chromatography is a required practical activity.

Notes

Thick-layer adsorption chromatography can be used as a purification technique for up to 100 mg of a sample. The purified, adsorbed product can be scraped off the plate, dissolved in a suitable solvent and recovered.

Column chromatography

This simple technique is an example of adsorption chromatography where the stationary phase is finely-divided alumina or silica gel contained in a vertical glass tube (the column). The moving phase (called the **eluent**) is usually an organic solvent.

A solution of the mixture is added to the top of the column, followed by enough fresh solvent to wash the components down the column; this process is referred to as **elution**.

The most strongly adsorbed components take the longest time to flow through the column. In general, the more polar the molecule, the greater the **retention time**. The eluent is either a pure solvent or a mixture of solvents chosen so that the different compounds can be separated effectively. Ideally, the components should elute one at a time from the column. Coloured components can be seen through the glass wall of the column as moving bands; fluorescent compounds can be visualised with the aid of an ultra-violet lamp. The solution emerging from the column (the **eluate**) is normally collected throughout the chromatographic separation as a series of fractions. Each fraction can be inspected separately and analysed for dissolved compounds.

Essential Notes

In a school laboratory, it is sometimes convenient to use an ordinary burette as a chromatography column.

Definition

*The **retention time** is the time each component remains in the column.*

Column chromatography provides a convenient way of separating and purifying individual organic compounds from mixtures. The whole process can be speeded up by forcing the solvent through the system under pressure (flash chromatography).

Gas chromatography (GC)

This analytical technique is a powerful method for the separation of mixtures of volatile compounds. The procedure uses a **carrier gas** (acting as eluent), such as helium or nitrogen, as the moving phase, and an inert powder coated with a film of a non-volatile liquid as the stationary phase. This coated powder is packed into a long, narrow-bore stainless steel or glass coiled tube (the column). Some chromatographs use very long, coiled capillary columns coated on the inside with the active film of non-volatile liquid.

Definition

*A **chromatograph** is the apparatus used for chromatographic separation of volatile components in a mixture.*

A continuous, steady flow of carrier gas passes through the column; the higher the flow rate the faster the analysis, but the lower the resolution.

The vaporised sample is injected into the entrance (head) of the column and the components are carried through the system and appear later, in sequence, at the exit. Each component has a characteristic retention time that depends on factors such as the nature of the stationary phase, the operating temperature, the flow rate of the carrier gas and the length of the column. Solubility in the non-volatile liquid phase is of major importance. A component that is highly soluble in the liquid phase takes considerably longer to elute from the column than one having a low solubility.

A detector is used to monitor the outlet stream from the column and this is linked to a recorder, so that each component appears as a peak on a chart. In conjunction with internal standards, retention times can be used to identify components, and peak areas are proportional to the amounts of different substances present.

GC can be used to measure minute quantities of chemicals and to distinguish between closely related groups of molecules. Very often, the chromatograph is connected to a mass spectrometer which operates as a detector capable of analysing in detail each separated component present in the mixture as it emerges from the column. This combination of techniques, abbreviated to GC-MS, is used in the testing of blood and urine samples for evidence of the use of illegal drugs.

Appendix

Summary of organic reactions in Collins Student Support Materials AS/A-Level year 1 - Organic and Relevant Physical Chemistry and this book

Alkanes

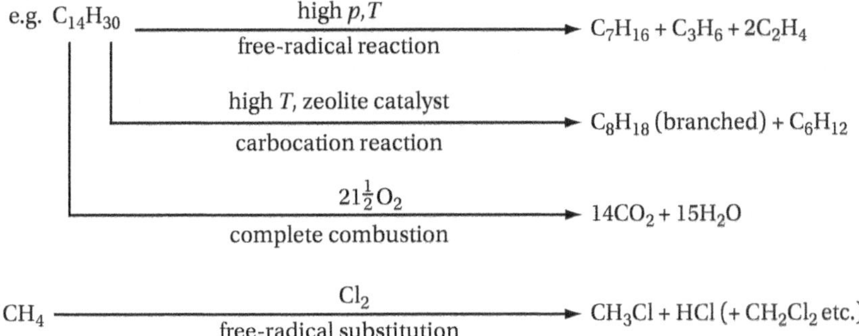

Alkenes

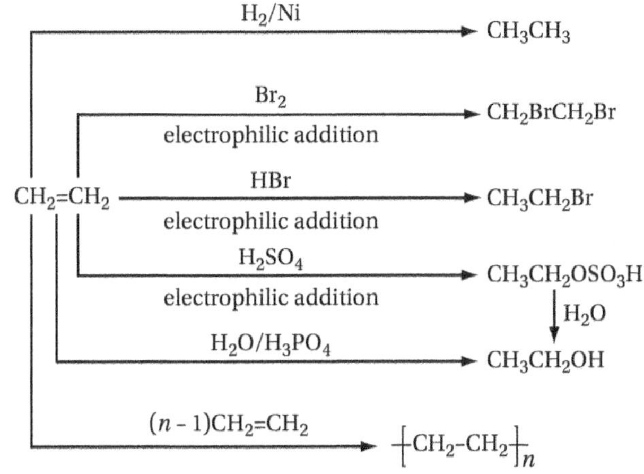

$CH_3CH{=}CH_2$:

- (1) H_2SO_4, (2) H_2O → $CH_3CH(OH)CH_3$ major product
- HBr → $CH_3CHBrCH_3$ major product
- $(n - 1)CH_3CH{=}CH_2$ → $\left[CH(CH_3)\text{—}CH_2\right]_n$

Halogenoalkanes

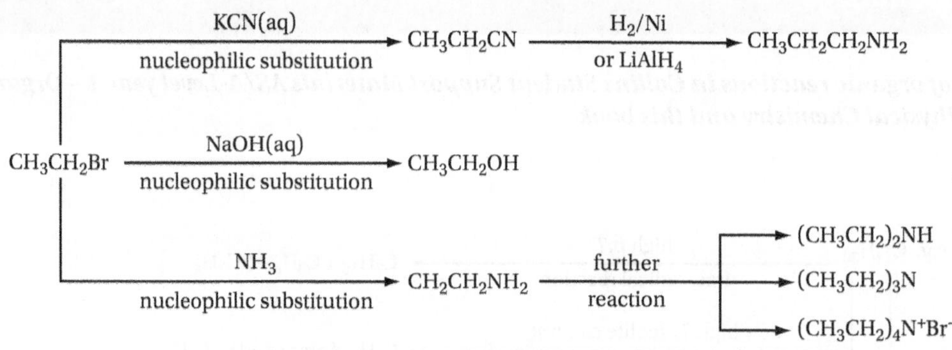

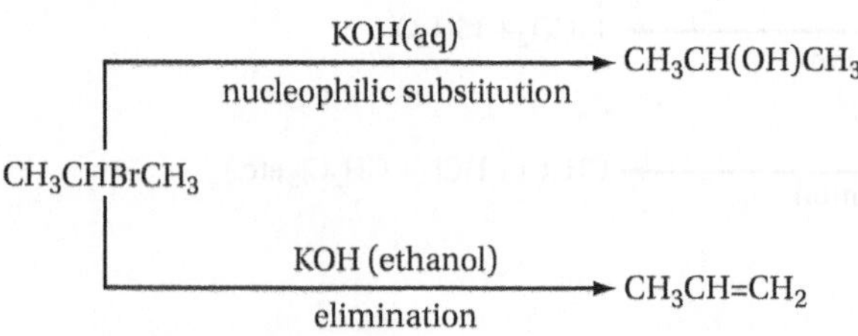

Alcohols

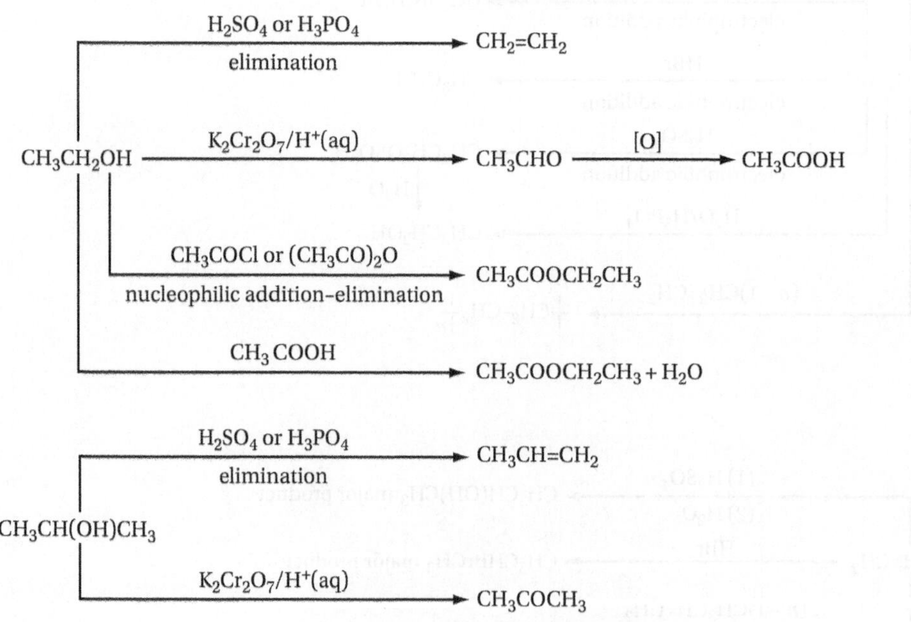

Aldehydes and ketones

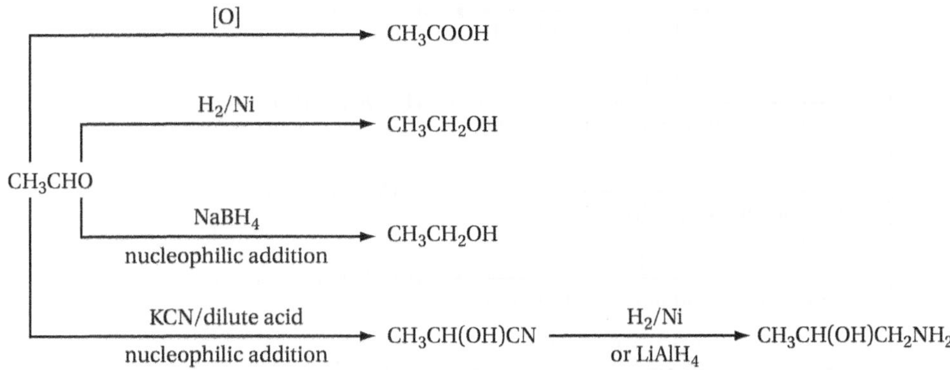

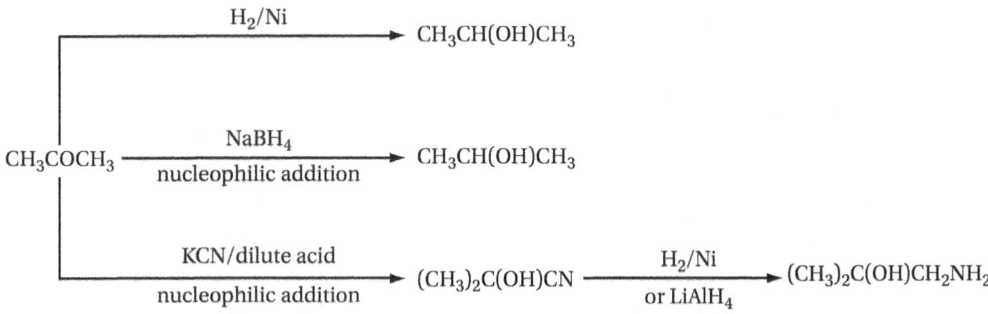

Carboxylic acids and esters

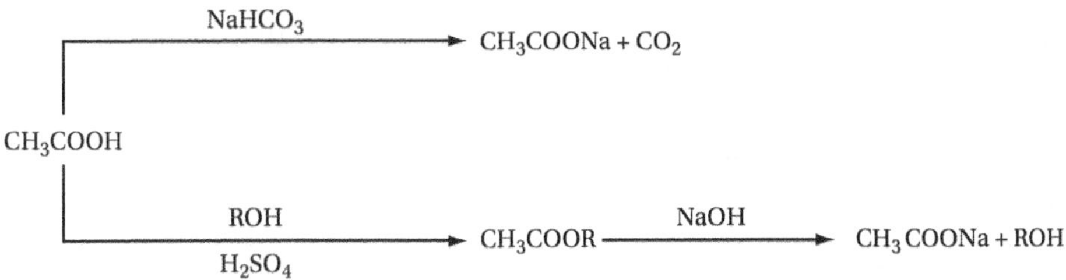

Acylation

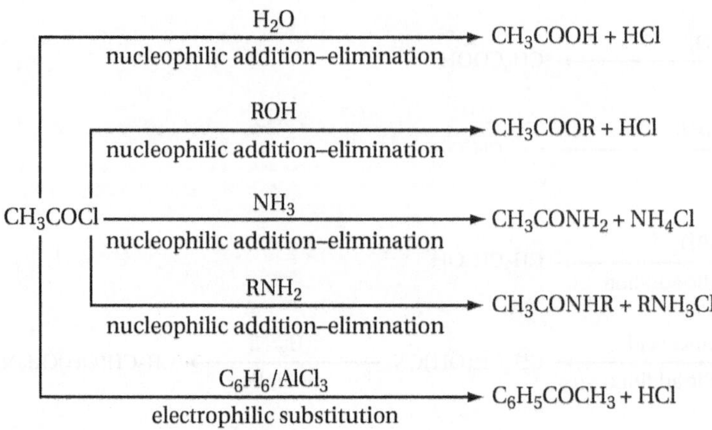

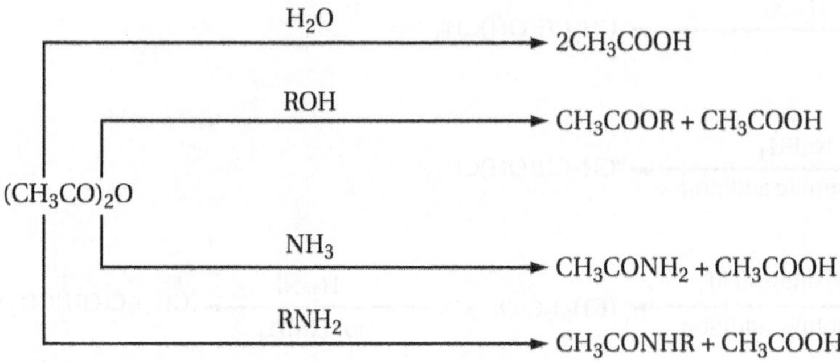

Aromatic chemistry

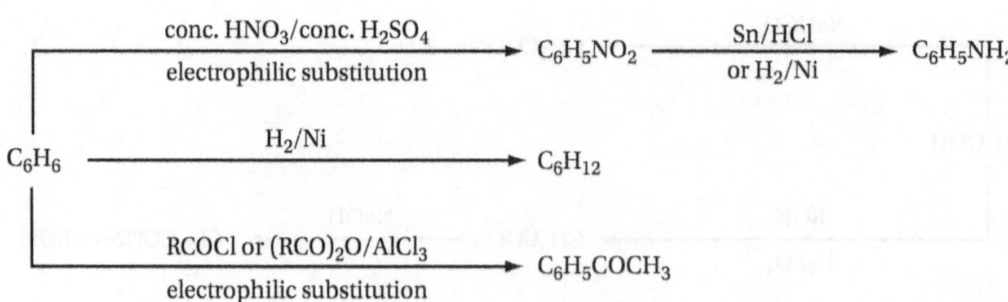

Practical and mathematical skills

In A-Level Paper 2, approximately 15% of marks will be allocated to the assessment of skills related to practical chemistry. A minimum of 20% of the marks will be allocated to assessing Level 2 mathematical skills. These practical and mathematical skills are likely to overlap.

In A-Level Paper 3, the practical and Level 2 mathematical skills that are assessed can be any of those covered in Books 1, 2, 3 and 4. Paper 3 will contain questions with a greater practical emphasis than the other two A-Level papers. Over all three A-Level papers, the practical skills assessed will be at least 15% of the marks and the Level 2 mathematical skills will be at least 20% of marks.

The required practical activities assessed in Paper 2 are:

- investigation of how the rate of a reaction changes with temperature
- distillation of a product from a reaction
- tests for alcohols, aldehydes, alkenes and carboxylic acids
- measuring the rate of a reaction by an initial rate method
- measuring the rate of a reaction by a continuous monitoring method
- preparation of a pure organic solid and testing its purity
- preparation of a pure organic liquid
- separation of species by thin-layer chromatography.

The practical skills assessed in the paper are as follows.

1. Independent thinking

Examination questions may require problem solving and the application of scientific knowledge and understanding in practical contexts. For example, a question could ask how an experiment could be carried out to determine how the rate of a reaction changes with temperature. Another example is a question that asks how the results of test-tube reactions can be used to distinguish between separate samples of an alcohol, an aldehyde, an alkene and a carboxylic acid.

2. Use and application of scientific methods and practices

This skill may be assessed by asking for critical comments on a given experimental method. Questions may ask how impurities can be removed from an organic solid and how the purity of the solid can be determined. Another question could ask how species can be separated by thin-layer chromatography and the certainty to which measured R_f values can be used to identify a component in a mixture. The factors that must be controlled and the measurements that must be made in experiments to determine the rate of a reaction must be known and understood.

3. Numeracy and the application of mathematical concepts in a practical context

There is some overlap between this skill and the use and the application of scientific methods and practices. Questions may require the plotting and use of concentration–time graphs to deduce rates of a reaction. Determination of reaction rate may be followed by questions asking how rate–concentration data or graphs are used to determine the order of reaction with respect to a given reactant.

4. Instruments and equipment

It will be necessary to know and understand how to set up glassware for the distillation of a product from a mixture, including the most appropriate method of heating the mixture. Questions may ask for a description of how a solid organic compound can be purified and how the melting point of the solid can be determined, and the apparatus required for each stage of this process. Knowledge and understanding of the equipment used to separate species by chromatography can be examined. Possible methods include thin-layer and column chromatography.

The mathematical skills assessed in this paper are:

1. Arithmetic and numerical computation

- **Recognise and make use of appropriate units in calculations.**

  All numerical answers should be given with the appropriate units. Question can be set that require the use of data to determine the order of a reaction with respect to a reagent. These calculations can include the requirement to deduce the units of the rate constant. Questions may require conversions between units, for example, changing volumes in cm^3 to dm^3 and time in minutes to seconds.

- **Recognise and use expressions in decimal and standard form.**

  When required, it will be necessary to express answers to an appropriate number of decimal places and to carry out calculations and express answers in ordinary or standard form. For example, calculations involving concentrations or rates of reaction may involve numbers in standard form and conversions between standard and ordinary form. The appropriate number of significant figures must be used in rate calculations.

- **Use ratios fractions and percentages.**

  Examples of this skill include the calculation of percentage yields, atom economies and the construction and/or balancing of equations using ratios.

- **Estimate results.**

  Estimations of this type could include how the rate of a reaction would be affected by a change in temperature.

- **Use of calculators.**

  The ability to use calculators to find power, exponential and logarithmic functions and to use these functions in the calculation of rate of reaction and activation energy.

2. **Handling data**

- **Use an appropriate number of significant figures.**

  Understand that a calculated result can only be reported to the limits of the least accurate measurement. For example, the measurement of temperature in experiments to determine activation energy and the measurement of time for an observed change in experiments to determine rate of reaction.

- **Find arithmetic means.**

  Examples may include the determination of the mean bond enthalpy from data for a given bond enthalpy in a range of compounds.

- **Identify uncertainties in measurements and when data are combined.**

  It will be necessary to demonstrate an ability to determine uncertainty when experimental time and temperature data are used to calculate an activation energy.

3. **Algebra**

- **Change the subject of an equation.**

  For example, when a rate equation is rearranged so that the rate constant for a reaction between two or more reagents is determined from the results of experiments with known concentrations of reagents.

- **Substitute numerical values into algebraic equations using appropriate units for physical quantities.**

  For example, using the Arrhenius equation to calculate the activation energy of a reaction from rate of reaction data obtained at different temperatures.

4. **Graphs**

- **Plot two variables from experimental data.**

  Examples of this skill include the plotting of concentration–time graphs. Questions can ask for tangents to be drawn to a curve and the slopes of these tangents used to determine the rates of reaction at different values of concentration of reactant. If rate of reaction is then plotted against an appropriate function of concentration, the order of a reaction can be determined.

5. **Geometry and trigonometry**

- **Visualise and represent 2D and 3D forms.**

  Questions may assess the ability to draw different forms of isomers and identify chiral centres from 2D and 3D representations. In addition, questions may ask for understanding of the symmetry of 2D and 3D shapes and identification of the types of stereoisomerism shown by molecules.

Practice exam-style questions

1 Phenyldiazonium chloride, $C_6H_5N_2Cl$, reacts with water to form phenol, C_6H_5OH, and nitrogen gas according to the following equation:

$$C_6H_5N_2Cl(aq) + H_2O(l) \rightarrow C_6H_5OH(aq) + N_2(g) + HCl(aq)$$

The structure of the diazonium ion is shown below:

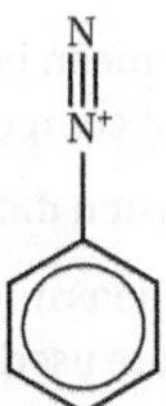

In a student's experiment to determine the order of the reaction, 50 cm^3 of a solution of concentration 0.080 mol dm^{-3} were heated at 323 K. The volume of nitrogen evolved over the course of 750 seconds was measured. This volume of nitrogen was used to calculate the concentration of the phenyldiazonium chloride solution. The graph of the concentration of phenyldiazonium chloride against time is shown below.

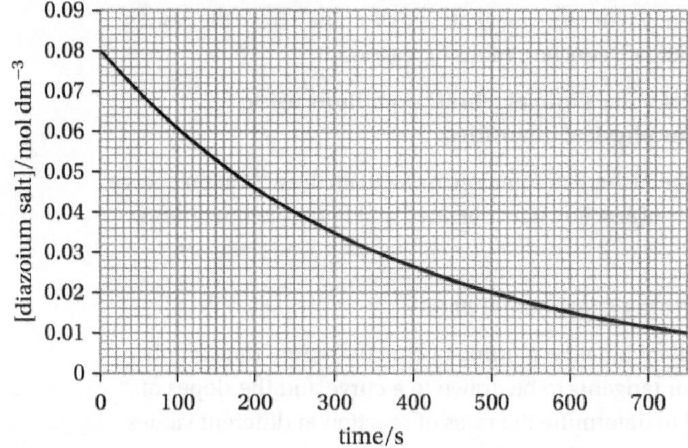

The gas constant, R, is 8.31 J K^{-1} mol^{-1}.

(a) The initial rate of the reaction is 2.24×10^{-4} mol dm^{-3} s^{-1}.

(i) Draw the gradient on the graph at 250 s. Use the gradient and the initial rate information to show that the reaction is first order with respect to the phenyldiazonium chloride. Show your working.

____

____

____ 3 marks

(ii) Use the initial rate to determine the value of the rate constant, *k*, at this temperature. Give your answer to an appropriate number of significant figures and give the units for the rate constant.

3 marks

(iii) The reaction is thought to occur in three steps. Outline a curly-arrow mechanism for the reaction. Label the rate-determining step and explain how the rate equation supports your choice.

6 marks

(b) Sketch a labelled diagram of the apparatus you could use to carry out the experiment.

3 marks

(c) Use the information provided to show that the total volume of nitrogen gas given off when all the diazonium salt had decomposed would be less than 100 cm^3 at 25 °C and 100 kPa.

____

____

____ 3 marks

(d) In a separate series of experiments, the decomposition of 3-methylphenyldiazonium chloride was investigated. The rate constants for the reaction were determined at different temperatures. The results are shown in Table 1.

Table 1

temperature/K	298	308	318	328	338
value of k	6.56×10^{-5}	5.01×10^{-4}	3.36×10^{-3}	2.01×10^{-2}	0.108

These results can be used to find the activation energy, E_a, of the reaction using the equation:

$k = Ae^{-E_a/RT}$

This equation can be re-arranged into the form

$\ln k = -E_a/RT + \ln A$

(i) Complete Table 2 below by calculating the missing values of $1/T$ and $\ln k$ and by writing your answers in the table.

Table 2

temperature/K	298	308	318	328	338
value of k	6.56×10^{-5}	5.01×10^{-4}	3.36×10^{-3}	2.01×10^{-2}	0.108
value of 1/T	3.36×10^{-3}	3.21×10^{-3}		3.05×10^{-3}	
ln k	-9.63	-7.60		-3.90	

2 marks

(ii) Label the axes below, plot a graph of $\ln k$ against $1/T$ and draw a straight line of best fit. 4 marks

(iii) Find the gradient of the graph and use it to find the value of the activation energy, E_a, for the decomposition of 3-methylphenyldiazonium chloride. Give the units for your answer.

4 marks

Total marks : 28

2 Propanone reacts with iodine in the presence of hydrogen ions to form iodopropanone. The reaction may be monitored using a colorimeter or visible spectrometer.

The equation for the reaction is:

$$CH_3COCH_3 + I_2 \rightarrow CH_3COCH_2I + HI$$

The absorbance can be used to calculate the concentrations of the reactants at given times.

A student carried out three experiments at 298 K to find the orders of the reaction with respect to iodine, hydrogen ions and propanone. The initial concentrations of each reactant are shown in Table 3 below.

Table 3

	Experiment 1	**Experiment 2**	**Experiment 3**
[propanone]/mol dm^{-3}	0.20	0.10	0.20
[H$^+$]/mol dm^{-3}	0.050	0.10	0.10
[I$_2$]/mol dm^{-3}	0.00104	0.00102	0.000995

A plot of concentration of iodine against time is shown for Experiment 1 and for Experiment 2. The results of Experiment 3 are given in Table 4.

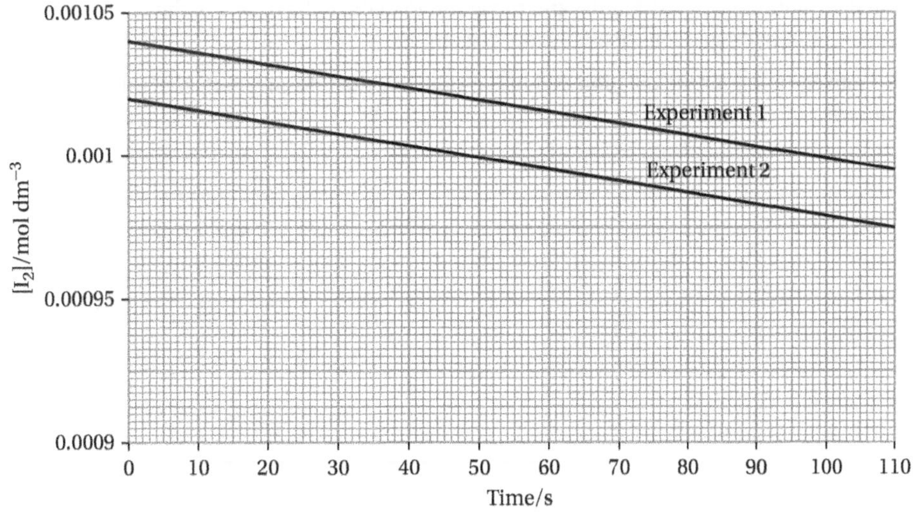

(a) **(i)** Suggest what properties of iodine allows the reaction to be monitored using these instrumental methods.

1 mark

(ii) Give two advantages of using a colorimeter to monitor a reaction compared with removing samples and titrating them.

2 marks

(b) The values of the concentration of iodine *versus* time for Experiment 3 are shown below in Table 4.

Table 4

Time/s	0	10	20	30	40	50
$[I_2]$/mol dm^{-3}	0.000995	0.000987	0.000981	0.000970	0.000962	0.000954
Time/s	60	70	80	90	100	110
$[I_2]$/mol dm^{-3}	0.000944	0.000938	0.000929	0.000921	0.000913	0.000905

Plot the points on the axes above and draw a best-fit straight line.

3 marks

(c) Calculate the rates of each experiment.

Use the information in Tables 3 and 4 and the graph to deduce the orders of the reaction with respect to iodine, propanone, and hydrogen ions. Explain your reasoning.

Write a rate equation for the reaction and find a value for the rate constant, *k*, at this temperature. Give units in your answer.

6 marks

(d) Suggest why the rate of the experiment might be expected to increase during the course of the reaction and why, in practice, the rate does not increase during the course of each run.

2 marks

Total marks: 14

3 Dinitrogen tetroxide, N_2O_4, decomposes on heating to form nitrogen dioxide according to the equilibrium below.

$$N_2O_4(g) \rightleftharpoons 2NO_2(g) \qquad \Delta H = +57 \text{ kJ mol}^{-1}$$

(a) At a given temperature, 0.304 mol of dinitrogen tetroxide is allowed to reach equilibrium in a 5.0 dm^3 flask. At equilibrium, 45% of the dinitrogen tetroxide has dissociated.

(i) Calculate the value of the equilibrium constant, K_c, at this temperature. Include units in your answer and give your answer to an appropriate number of significant figures.

6 marks

(ii) Deduce the value of K_c for the equilibrium below at the same temperature. Suggest units for this new equilibrium constant.

$$\tfrac{1}{2}N_2O_4(g) \rightleftharpoons NO_2(g)$$

2 marks

(b) State and explain the effect of increasing the temperature on:

(i) the position of equilibrium.

2 marks

(ii) the value of K_c.

2 marks

Total marks: 12

4 Benzoic acid may be prepared by the hydrolysis of methyl benzoate.

$$C_6H_5COOCH_3 + H_2O \longrightarrow C_6H_5COOH + CH_3OH$$

A student prepared a sample of benzoic acid as follows.

A reaction mixture of aqueous sodium hydroxide and methyl benzoate was boiled under reflux until all the oily droplets had disappeared. The mixture was cooled to room temperature. Dilute sulfuric acid was added until there was no further precipitate. The precipitate was filtered off and recrystallised.

(a) Explain how and why the reaction mixture is boiled under reflux.

3 marks

(b) Explain how the precipitate can be made more pure by recrystallisation.

3 marks

(c) In a typical experiment, the yield of purified crystals of benzoic acid obtained from 0.050 mol of methyl benzoate is 5.2 g. Calculate the percentage yield of benzoic acid. Give your answer to an appropriate number of significant figures.

2 marks

(d) The identity and purity of the product can be investigated by a number of experimental techniques. Explain how each of the following can be used to check the identity of the benzoic acid product:

(i) the melting point of the crystals.

2 marks

(ii) the fingerprint region of the infra-red.

2 marks

(e) ^{1}H NMR spectroscopy and thin-layer chromatography, TLC, can be used to monitor the course of the reaction.

(i) Explain how TLC can be used to show that the methyl benzoate has completely reacted.

2 marks

(ii) During the hydrolysis reaction, small samples are removed from the mixture and acidified. A ^{1}H NMR spectrum is obtained on this organic sample. Over the course of the hydrolysis, the singlet peak at 4.1 p.p.m. integrating for three protons disappears and a singlet peak at 10.5 p.p.m. integrating for one proton appears. Assign these peaks and explain why the peaks are not split.

3 marks

(f) A student determines the purity of the benzoic acid by titration.

In the student's analysis, 1.27 g of benzoic acid crystals are reacted with 25.0 cm^3 of 1.60 mol dm^{-3} sodium hydroxide solution. The mixture is transferred into a 250.0 cm^3 graduated flask and 25.0 cm^3 portions are titrated with 0.0500 mol dm^{-3} sulfuric acid solution. The results are shown below in Table 5.

(i) Calculate the percentage purity of the benzoic acid, assuming that any impurities in the benzoic acid do not react with either sodium hydroxide or sulfuric acid. Include in your answer equations for the reaction of sodium hydroxide with sulfuric acid and with benzoic acid.

Table 5

Titration	1	2	3	4
Final/cm^3	32.7	30.15	31.05	31.40
Initial/cm^3	2.5	0.10	1.25	1.60

11 marks

(ii) Deduce the percentage uncertainty in the use of the burette.

1 mark

(iii) A student suggested that the percentage uncertainty in this determination was greatest in the use of the graduated flask. Use the data below together with your answer to part (f)(ii) to decide whether or not the student was correct. Suggest how the percentage uncertainty in the measurement with the greatest uncertainty could be reduced.

The mass of benzoic acid was measured with a balance with an uncertainty of ± 0.01 g.

The volume of sodium hydroxide was measured with a pipette with a total uncertainty for two measurements of ± 0.12 cm^3.

The graduated flask has an uncertainty of ± 0.6 cm^3.

3 marks

Total marks: 32

5 (a) Name the type of stereoisomerism shown by 2-hydroxypropanoic (*lactic*) acid, $CH_3CH(OH)COOH$. Identify the structural feature of the molecule that permits the existence of two isomers. With the aid of diagrams, illustrate the structural relationship between these isomers. Explain how it is possible to distinguish between the two isomers.

Type of stereoisomerism 1 mark

Structural feature 1 mark

Isomer 1 *Isomer 2* 2 marks

Explanation

2 marks

(b) Explain, with reference to the isomers of lactic acid, the meaning of the terms *enantiomer* and *racemate.*

Enantiomer

Racemate ____

____

____ 4 marks

(c) One form of an enzyme, lactate dehydrogenase, is active for only one of the two enantiomers above.

Explain how the three-dimensional structure of the enzyme allows it to be specific to this one enantiomer.

____

____

____

____ 3 marks

(d) The three-dimensional shape of an enzyme is formed from three types of structure: *primary structure, secondary structure* and *tertiary structure*. Explain what is meant by each of these terms.

Primary structure ____

____

____

____ 2 marks

Secondary structure ____

____

____

____ 2 marks

Tertiary structure ____

____

____

____ 2 marks

(e) Identify the features of the protein shown in schematic form below that have been arrowed in the diagram.

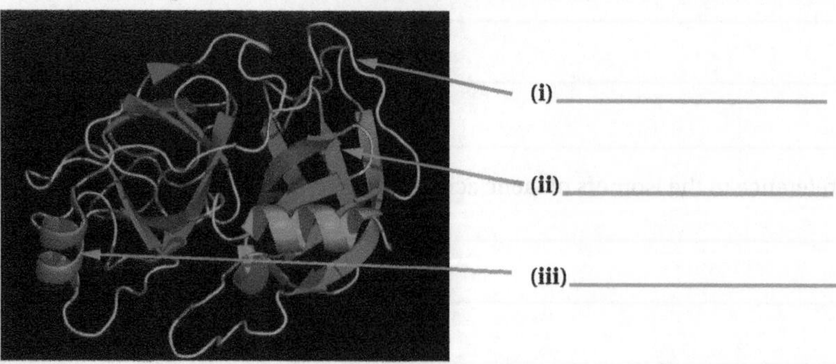

(i) ____________________

(ii) ____________________

(iii) ____________________

3 marks

Total marks: 22

6 Polypeptides can be converted into their component amino acids by hydrolysis.

(a) Explain what is meant by *hydrolysis*, and state the conditions under which a polypeptide can be hydrolysed.

meaning of hydrolysis ____

____

conditions ____

____ 2 marks

(b) A tripeptide is hydrolysed and the mixture of amino acids is analysed by TLC. The chromatogram is shown below.

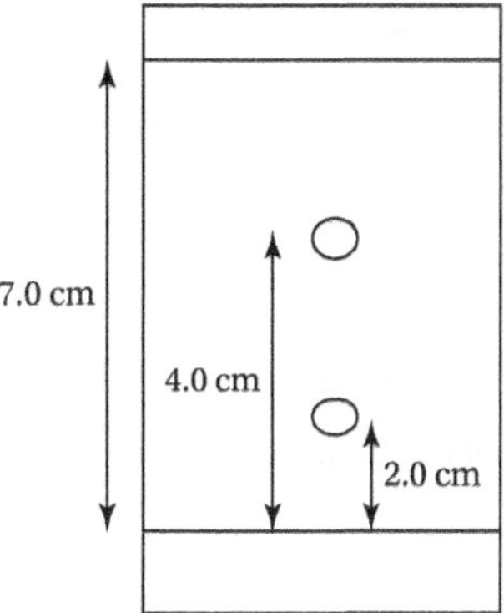

(i) A table of R_f values of amino acids is given below.

Amino acid	R_f value
alanine	0.39
glutamine	0.17
glycine	0.28
histidine	0.10
lysine	0.18
methionine	0.57
phenylalanine	0.69
serine	0.25
tyrosine	0.48

Identify the amino acids in the mixture.

____

____ 2 marks

(ii) Explain why R_f values are approximate but still allow identification of amino acids in this experiment.

____

____ 2 marks

(iii) State two methods which could be used to make the spots of the amino acids visible.

____

____ 2 marks

(c) Explain why the melting point of (+)-2-aminopropanoic acid (*alanine*), $CH_3CH(NH_2)COOH$ (314 °C), is much higher than that of (+)-2-hydroxypropanoic acid (53 °C).

____

____

____

____

____

____ 4 marks

(d) When an aqueous solution of alanine is electrolysed at low pH, an organic species is attracted to the cathode. At high pH, a different organic species is attracted to the anode. Deduce the structure of each of the moving species.

Species moving to cathode at low pH *Species moving to anode at high pH* 2 marks

(e) Cysteine has an important role in the 3-dimensional structure of a protein.

(i) Draw the structure of the dipeptide formed by two molecules of cysteine.

Dipeptide ____ 2 marks

(ii) Molecules of cysteine in different parts of a protein can form a different type of covalent bond from that in the peptide bond. Identify the feature of a cysteine molecule that allows this to happen. Draw the bond that is formed and suggest the type of reaction which occurs when two cysteine residues in a polypeptide form this bond.

Feature of cysteine ____________________

Bond ____________________

Type of reaction ____________________ 3 marks

(f) Using RNH_2 to represent alanine, write an equation for the reaction between alanine and ethanoyl chloride.

Name and outline the mechanism of the reaction.

Equation ____________________

Name of mechanism ____________________

Mechanism

6 marks

Total marks: 25

7 Aldehydes and ketones, such as pentan-2-one, react with hydrogen cyanide.

(a) Write an equation for the reaction between pentan-2-one and hydrogen cyanide. Name and outline the mechanism of the reaction. Give the IUPAC name of the product.

Equation ____________________

Name of mechanism ____________________

Mechanism

Name of product __________ 7 marks

(b) Explain how this reaction is carried out so that hazards are minimised.

____

__________ 1 mark

(c) State the type of stereoisomerism shown by the product and explain why the product obtained in the reaction in part (a) is formed as an optically-inactive mixture.

Type of stereoisomerism __________

Explanation __________

____

__________ 3 marks

Total Marks: 11

8 Butanal and butanone can readily be distinguished using ^{1}H NMR spectroscopy but not as easily by other spectroscopic techniques.

Explain how the number of peaks in the ^{1}H NMR spectrum and their integration ratio would allow the isomers to be distinguished.

Discuss whether the infra-red spectrum above 1600 cm^{-1}, the precise molecular mass and the number of peaks and their chemical shifts in the ^{13}C NMR spectrum would allow them to be distinguished.

6 marks

9 Tallow, obtained from beef fat, can be used to make soap and also biodiesel. A representative structure of tallow is shown below.

$$\begin{array}{l} CH_2OOC(CH_2)_{14}CH_3 \\ | \\ CHOOC(CH_2)_{16}CH_3 \\ | \\ CH_2OOC(CH_2)_7CH{=}CH(CH_2)_7CH_3 \end{array}$$

(a) What type of process is used to convert a fat of this type into a soap? Give the reagent that would be used for this purpose. Give the IUPAC name of the neutral organic by-product of this reaction.

Type of process ______

Reagent ______

IUPAC name of organic by-product ______ 3 marks

(b) **(i)** Unsaturated groups present in oils and fats have the *Z*-configuration in most cases. Using RCH=CHR, draw a structure to show what is meant by a *Z*-configuration. Name the alternative type of configuration.

Structure ______

Alternative type ______ 2 marks

(ii) Suggest a reagent system that can be used to convert an alkene into a saturated group.

______ 2 marks

(c) The formation of biodiesel from oils and fats involves heating them with an excess of a simple alcohol. Name this type of process and give the structure of one of the esters present in biodiesel that is obtained by heating tallow with an excess of methanol. Explain briefly why it is necessary for an excess of the alcohol to be present.

Type of process ____________________

Structure of ester ____________________

Explanation ____________________

____________________ 4 marks

(d) Terylene is made industrially by heating together ethane-1,2-diol and dimethyl benzene-1,4-dicarboxylate. Gaseous methanol is formed as a by-product. Draw the repeating unit of Terylene and explain why this polymer, like biodiesel, is biodegradable.

Repeating unit

Explanation ____________________

____________________ 4 marks

Total marks: 15

10 The local anaesthetic benzocaine (**D**) can be made by the following sequence.

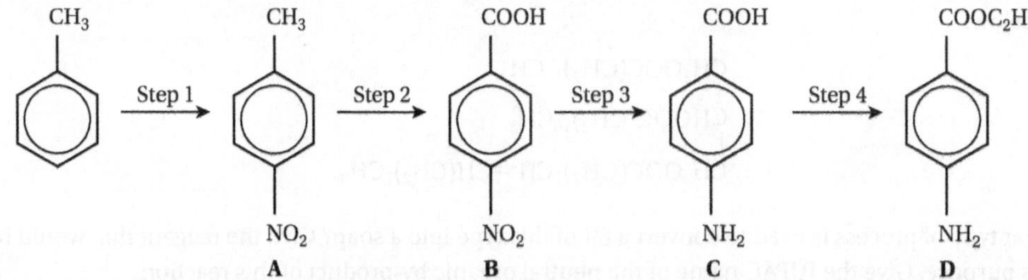

(a) In *Step 1*, methylbenzene is treated with a mixture of two reagents. Identify the two reagents and write an equation showing the formation of the reactive species involved. State the type of reaction taking place in the conversion of methylbenzene into **A**, name and outline the mechanism for the reaction of the reactive species with methylbenzene.

Reagent 1 ____________________

Reagent 2 ____________________

Equation ____________________

Type of reaction ____________________

Name of mechanism ____________________

Mechanism

7 marks

(b) The product obtained from *Step 1* is a liquid and contains compound **A** contaminated with other isomers.

When pure, compound **A** is a crystalline solid, m.p. 51 °C. Describe briefly how chromatography could be used to obtain a pure sample of crystalline **A** from 5 g of the impure liquid mixture.

5 marks

(c) Name the type of reaction occurring in *Step 2*.

1 mark

(d) Name the type of reaction taking place in *Step 3* and give a suitable reagent or mixture of reagents for this conversion.

Type of reaction

Reagent(s) 3 marks

(e) Compound **C** contains a small amount of unreacted compound **B**. Both compound **C** and compound **B** are solids which are sparingly soluble in water. Suggest a way of obtaining compound **C** as a pure sample and explain the basis of your method.

Method of separation ____________________

Explanation ____________________

____

____

____

____________________ 4 marks

(f) Name the type of reaction occurring in *Step 4* and identify the reagents needed for this conversion. Give the IUPAC name of compound **D**.

Type of reaction ____________________

Reagents ____________________

IUPAC Name ____________________ 4 marks

Total marks: 24

11 (a) Write an ionic equation to illustrate how the primary amine RNH_2 functions as a Brønsted–Lowry base. Explain why ethylamine, $C_2H_5NH_2$, is a stronger base than ammonia, whereas phenylamine, $C_6H_5NH_2$, is less basic than ammonia.

Equation ____________________

Explanation ____________________

____

____

____

____________________ 5 marks

(b) Give the structure and name the type of compound formed when R_2NH is heated with a large excess of chloromethane. Give one use of the product obtained when R is a very long alkyl group.

Structure ____________________

Type of compound ____________________

Use ____________________ 4 marks

(c) Use the Data Sheet to help you answer this part of the question.

Amines are found in the four bases in DNA.

A five-base sequence in one strand of DNA is found to be -A-T-T-G-G-.

(i) Explain how cisplatin would prevent replication of this strand of DNA.

Explanation ____________________

____

____

____________________ 3 marks

(ii) Give the five-base sequence in the complementary strand of DNA.

____________________ 1 mark

(iii) Name the type of interaction between the base-pairs in the two strands of the DNA.

____________________ 1 mark

Total marks: 14

12 The —CONH— unit occurs in various types of organic compounds, ranging from simple amides and cyclic structures to synthetic polymers and complex proteins.

(a) Caprolactam, the cyclic amide of 6-aminohexanoic acid, is the monomer used in a ring-opening polymerisation reaction to manufacture nylon-6.

(i) Draw the structure of caprolactam and give the repeating unit of nylon-6.

Structure of caprolactam ____________________

Repeating unit of nylon-6 ____________________ 2 marks

(ii) Deduce the structure of the product obtained when caprolactam is hydrolysed with aqueous sodium hydroxide.

1 mark

(b) The aromatic polymer Kevlar is obtained from the reaction between compounds **E** and **F**.

(i) Give the IUPAC names of **E** and **F**.

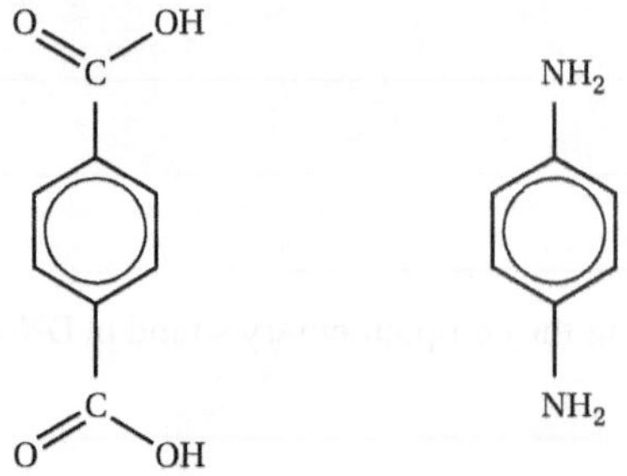

E **F**

IUPAC name of **E** ______

IUPAC name of **F** ______ 2 marks

(ii) State the type of condensation polymer.

______ 1 mark

(iii) Draw the repeating unit of Kevlar and explain why this polymer has a sheet-like structure.

Repeating unit of Kevlar

Explanation ______

______ 3 marks

(c) The hydrolysis of a protein molecule, such as insulin, yields a mixture of amino acids. Suggest a laboratory technique suitable for the separation of the components present in a mixture of amino acids. Suggest a means of identification of the various components present in such a mixture.

Laboratory technique ______

Means of identification ______ 2 marks

(d) A dipeptide is formed by the combination of two amino acids. Deduce the number of different dipeptides that can be produced by a combination of two of twenty naturally-occurring amino acids.

Number of possible dipeptides ______ 1 mark

Total marks: 12

13 Various polymers have monomers which can be represented by $XCH{=}CH_2$. Examples include poly(propene) ($X = CH_3$) and poly(chloroethene) ($X = Cl$). Typical uses of these products include tubing and food containers. Used products could end up on landfill sites, be recycled or incinerated.

(a) State the type of polymer formed by the monomers given above and draw the repeating unit of poly(chloroethene).

Type of polymer ____________________

Repeating unit ____________________ 2 marks

(b) Explain briefly why neither of these polymers is biodegradable.

____

____

____________________ 2 marks

(c) Suggest one advantage and one disadvantage of recycling poly(propene).

Advantage ____________________

Disadvantage ____________________ 2 marks

(d) Suggest one advantage and one disadvantage of incinerating poly(chloroethene).

Advantage ____________________

Disadvantage ____________________ 2 marks

Total marks: 8

14 A mixture of two organic liquids could not be separated efficiently using fractional distillation. Use of a different laboratory technique permitted the isolation of compound **G**, b.p. 202 °C, and compound **H**, b.p. 205 °C.

Compound **G** has the molecular formula, C_8H_8O. A dominant peak appears in the infra-red spectrum of **G** at 1690 cm^{-1}. One peak in the ^1H NMR spectrum of **G** is a three-proton singlet at $\delta = 2.60$ ppm and one of the peaks present in the ^{13}C NMR spectrum has a strong signal at $\delta = 198$ ppm. Compound **G** remains unchanged when heated with acidified potassium dichromate(VI) solution.

Compound **H** has the molecular formula C_7H_8O. An intense peak appears in the infra-red spectrum of **H** at 3352 cm^{-1}. One peak in the ^1H NMR spectrum of **H** is a two-proton singlet at $\delta = 4.50$ ppm. Compound **H** is oxidised to benzenecarboxylic acid when heated with acidified potassium dichromate(VI) solution.

In the ^{13}C NMR spectrum, both **G** and **H** have four peaks between $\delta = 110$ and 160 ppm.

(a) Suggest which technique was used to separate compounds **G** and **H**.

____________________ 1 mark

(b) What conclusion can be drawn from the fact that both **G** and **H** have **four** peaks between $\delta = 110$ and 160 ppm in their ^{13}C NMR spectra?

____

____

____________________ 2 marks

(c) For compound **G**, identify the groups associated with the following peaks.

1690 cm^{-1} in the infra-red spectrum ______

$\delta = 2.60$ ppm in the ^1H NMR spectrum ______

$\delta = 198$ ppm in the ^{13}C NMR spectrum ______ 3 marks

(d) Deduce the structure of compound **G**.

______ 1 mark

(e) For compound **H**, identify the groups associated with the following peaks.

3352 cm^{-1} in the infra-red spectrum ______

$\delta = 4.50$ ppm in the ^1H NMR spectrum ______ 2 marks

(f) Deduce the structure of compound **H**.

______ 1 mark

(g) Explain how the behaviour of compounds **G** and **H** towards acidified potassium dichromate(VI) solution relates to the structures of these compounds.

______

______

______

______ 2 marks

Total marks: 12

15

Route 1

step 1 → step 2 → NH_2

K ↑ step 5

step 3 → step 4 → J

Route 2

The scheme shows how methylcyclopentane and cyclopentene can be converted into aminomethylcyclopentane, **K**.

(a) Give reagents and conditions for steps 1 and 2.

4 marks

(b) Hydrogen bromide reacts with cyclopentene in step 3. Name and outline a mechanism for this step.

5 marks

(c) Give reagents and conditions for step 4. Name and outline a mechanism for this step.

5 marks

(d) Draw the structure of the intermediate, **J**, and give the reagent(s) and conditions needed to convert **J** into the product, **K**.

3 marks

(e) The percentage yields for each step are shown in the Table 6. Calculate the overall yield for each route to form Compound **K**. Give reasons why steps 1 and 2 each have low yields.

Table 6

Step	1	2	3	4	5
Percentage yield	20	25	90	85	80

6 marks

Total marks: 23

16 Give a reagent which could be used in a simple chemical test to distinguish between the following pairs. Give the observations for each compound.

(a) cyclopentane and pent-2-ene.

Reagent

Observation for cyclopentane

Observation for pent-2-ene 3 marks

(b) $CH_3CH_2COOCH_3$ and $CH_3OCH_2CH_2CHO$.

Reagent

Observation for $CH_3CH_2COOCH_3$

Observation for $CH_3OCH_2CH_2CHO$ 3 marks

(c) $CH_3COOCH_2CH_3$ and $CH_3CH_2CH_2COOH$

Reagent

Observation for $CH_3COOCH_2CH_3$

Observation for $CH_3CH_2CH_2COOH$ 3 marks

(d) pentan-2-ol and 2-methylbutan-2-ol

Reagent

Observation for pentan-2-ol

Observation for 2-methylbutan-2-ol 3 marks

(e) propanoyl chloride and 1-chloropropane.

Reagent

Observation for propanoyl chloride

Observation for 1-chloropropane 3 marks

Total marks: 15

17 Draw structures for each of the following pairs of isomers. Use the representation you are directed to draw for each part.

(a) **L** and **M** have the molecular formula C_6H_{12}. **L** has three peaks in its ^{13}C NMR, an absorption at 1650 cm^{-1} in the infra-red and has *E–Z* isomers. **M** has one peak in the ^{13}C NMR and has no absorption between 1600 and 1800 cm^{-1} in the infra-red.

Draw skeletal formulae for **L** and **M**

L ____________________ **M** ____________________

2 marks

(b) **P** and **Q** have molecular formula $C_3H_6Br_2$. **P** has stereoisomers but **Q** does not. **P** has three peaks in the ^1H NMR, two of which are doublet peaks, whereas **Q** has two peaks, one of which is a triplet.

Draw displayed formulae for **P** and **Q**.

P ____________________ **Q** ____________________

2 marks

(c) **R** and **S** have molecular formula $C_5H_{10}O_2$. **R** has a strong absorption in the infra-red at 1740 cm^{-1} and four peaks in its ^1H NMR, two quartets and two triplets. The two quartets have chemical shift $\delta = 4.1$ and 2.5 ppm. **S** has a strong absorption in the infra-red at 1705 cm^{-1} and a broad absorption between 2500 and 3000 cm^{-1}. **S** has two singlet peaks in the ^1H NMR at chemical shift $\delta = 10.4$ and 1.1 ppm integrating for one and nine protons, respectively.

Draw skeletal formulae for **R** and **S**.

R ____________________ **S** ____________________

2 marks

Total marks: 6

Multiple choice questions

1 In an experiment to investigate an esterification reaction, an equal number of moles of methanol and propanoic acid were allowed to reach equilibrium. At equilibrium, 0.100 mol of methyl propanoate was formed. The equilibrium constant, K_c, was found to be 4.10. The original amount of methanol, in moles, was:

A 0.0494

B 0.102

C 0.149

D 0.256

2 The rate equation for a reaction is found to be

rate = $k[\mathbf{T}]^2[\mathbf{U}]$

At 298 K, the rate is found to be 1.27×10^{-3} mol dm^{-3} s^{-1} when the concentration of **T** is 0.15 mol dm^{-3} and **U** is 0.090 mol dm^{-3}.

The value of the rate constant, *k*, is

A 2.6×10^{-6}

B 1.7×10^{-5}

C 0.094

D 0.63

3 Which of the following types of mechanism would compound **V** not undergo?

V

A electrophilic addition

B electrophilic substitution

C nucleophilic addition

D nucleophilic substitution

4 In which of the following are the substances arranged in the correct order for the named physical characteristic?

A basicity: ethylamine < ammonia < phenylamine

B boiling point: pentane < methylbutane < dimethylpropane

C rate of reaction with aqueous NaOH: 1-fluorobutane < 1-chlorobutane < 1-bromobutane

D rate of hydrolysis: ethanoyl chloride < chloroethane < chlorobenzene

Answers

Question	Answer		Marks
1 (a) (i)	[diazoium salt]/mol dm^{-3} vs time/s gradient drawn on graph	(1)	
	gradient = (0.068 – 0.012)/(500 – 0) = 1.12×10^{-4} mol dm^{-3} s^{-1}	(1)	
	rate at concentration of diazonium chloride of 0.04 mol dm^{-3} is half the initial rate at concentration 0.08 mol dm^{-3} so first order in diazonium salt	(1)	3
1 (a) (ii)	rate = k[diazonium chloride]		
	$\therefore$ k = rate/[diazonium chloride]	(1)	
	$\therefore$ $k = 2.24 \times 10^{-4}/0.08 = 2.775 \times 10^{-3} = 2.8 \times 10^{-3}$ (2 sig. figs.)	(1)	
	units of k = s^{-1}	(1)	3
1 (a) (iii)	step 1 step 2 step 3 (1) mark for curly arrow to N$^+$ in step 1 (1) mark for phenylcarbocation structure (1) mark for curly arrow from lone pair on O to + charge on phenyl ring (1) mark for curly arrow from an O–H bond to O (not + charge)		

Question	Answer		Marks
	Explanation: only benzenediazonium chloride is in rate equation rate determining step must only include benzenediazonium chloride and not water so step 1 is rate determining step	(1) (1)	6
1 (b)	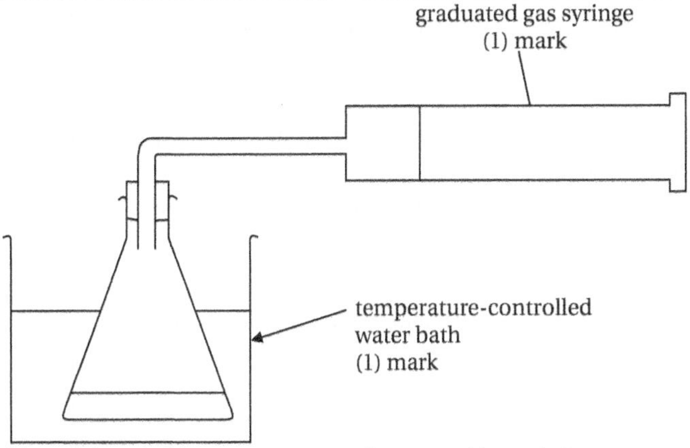 graduated gas syringe (1) mark temperature-controlled water bath (1) mark (1) mark for suitable apparatus correctly set up with no obvious gaps (1) mark for suitable way of controlling temperature (1) mark for way of measuring volume of gas, *e.g.* gas syringe or inverted burette		3
1(c)	initial moles of diazonium salt = moles of N_2 formed = 0.0040 volume = $nRT/p = 0.0040 \times 8.31 \times 298/100000\ m^3$ volume = $9.9055 \times 10^{-5}\ m^3 = 99\ cm^3$ (which is less than $100\ cm^3$)	(1) (1) (1)	3
1 (d) (i)	**temperature/K** 318 338 **value of *k*** 3.36×10^{-3} 0.108 **value of 1/*T*** $\mathbf{3.14 \times 10^{-3}}$ $\mathbf{2.96 \times 10^{-3}}$ **ln *k*** **−5.70** **−2.22** (1) mark for correct values of $1/T$ (1) mark for correct values of ln k		2
1 (d) (ii)	y-axis labelled ln k and x-axis labelled $1/T$ with a scale that covers more than half of each axis all points plotted correctly straight line of best fit drawn ignoring point at $[3.21 \times 10^{-3}; -7.60]$	(1) (1) (1) (1)	4
1 (d) (iii)	value of gradient = $[-9.36 - (-2.22)]/(0.00336 - 0.00296) = -17850$ gradient = $-E_a/R$ $\therefore -E_a = -17850 \times R = -17850 \times 8.31 = -148\,333.5\ J\ mol^{-1}$ $\therefore E_a = 150\ kJ\ mol^{-1}$	(1) (1) (1) (1)	4
			Total 28
2 (a) (i)	iodine solutions are coloured	(1)	1
2 (a) (ii)	smaller sample needed continuous monitoring is possible (if connected to computer or chart recorder)/ many more points can be recorded	(1) (1)	2
2 (b)	all points plotted correctly line of best fit drawn ignoring points at [20,0.000981] and [60.0.000944]	(1) (1) (1)	3

Question	Answer	Marks
2 (c)	**This answer is marked using levels of response.**	
	Level 3: 5–6 marks All parts are covered and the explanation of each part is generally correct and virtually complete. Answer communicates the whole process coherently and shows a logical progression from part 1 and part 2 to overall explanation.	
	Level 2: 3–4 marks All parts are covered but the explanation of each part may be incomplete OR two parts are covered and the explanations are virtually complete. Answer is mainly coherent and shows a progression. Some statements may be out of order and incomplete.	
	Level 1: 1–2 marks Two parts are covered but the explanation of each part may be incomplete and contain inaccuracies OR only one part is covered but the explanation is mainly correct and is virtually complete. Answer includes some isolated statements but there is no attempt to present them in a logical order or show confused reasoning.	
	Level 0: 0 marks	
	Part 1: rates of reaction: rate of experiment 1 = $(0.00104 - 0.00100)/(100 - 0) = 4.0 \times 10^{-7}$ mol dm^{-3} s^{-1} rate of experiment 2 = $(0.00102 - 0.00098)/(100 - 0) = 4.0 \times 10^{-7}$ mol dm^{-3} s^{-1} rate of experiment 3 = $(0.000995 - 0.000913)/(100 - 0) = 8.0 \times 10^{-7}$ mol dm^{-3} s^{-1} *Part 2: orders of reaction:* plot of $[I_2]$ against time is straight line ∴ 0 order in iodine as rate does not change as concentration of iodine changes comparing experiments 1 and 3, $[H^+]$ doubles and the rate doubles ∴ first order in $[H^+]$ comparing experiments 2 and 3, [propanone] doubles and the rate doubles ∴ first order in [propanone] *Part 3: calculation of rate constant and units:* Rate equation is rate = $k[H^+]$[propanone] ∴ k = rate/{$[H^+]$[propanone]} ∴ $k = 4.0 \times 10^{-7}/[0.050 \times 0.20]$ $= 4.0 \times 10^{-5}$ mol^{-1} dm^{3} s^{-1}	6
2 (d)	HI is a product of the reaction and is acidic so $[H^+]$ increases (1) change in $[H^+]$ is too small (1)	2
		Total 14
3 (a) (i)	amount of dinitrogen tetroxide remaining at equilibrium $= 0.304 \times 55/100$ or $(0.304 - 0.304 \times 45/100)$ $= 0.1672$ mol (1) amount of NO_2 at equilibrium $= 0.304 \times 45/100 \times 2 = 0.2736$ mol (1) concentration of N_2O_4 at equilibrium $= 0.1672/5 = 0.03344$ mol dm^{-3} concentration of NO_2 at equilibrium $= 0.1672/5 = 0.05472$ mol dm^{-3} (1) for both values $K_c = [NO_2]_2/[N_2O_4]$ (1) $= 0.05472^2/0.03344 = 0.0895$ $= 0.090$ (2 sig. fig) (1) units of K_c: mol dm^{-3} (1)	6

Question	Answer	Marks
3 (a) (ii)	K_c of new equilibrium = $[NO_2]/[N_2O_4]^{1/2}$ $\therefore K_c = 0.30$ (square root of answer to part 3(a)(i)) (1) units: $mol^{1/2}\ dm^{-3/2}$ (1)	2
3 (b) (i)	reaction is endothermic since ΔH is positive (+57 kJ mol^{-3}) (1) position of equilibrium moves to the right (1)	2
3 (b) (ii)	K_c increases (1) position of equilibrium moves to right (1)	2
		Total 12
4 (a)	to ensure all the ester is hydrolysed and the benzoic acid converted to the benzoate ion (1) to condense volatile substances (1) to prevent volatile substances being lost (1)	3
4 (b)	use the minimum of solvent to dissolve the benzoic acid (1) solvent must be boiling (1) mixture allowed to cool until crystallisation is complete (1)	3
4 (c)	amount of benzoic acid = mass/M_r = 5.2/122.0 = 0.0426 mol (1) 1:1 ratio so maximum yield of benzoic acid is 0.050 mol % yield = 0.0426/0.050 × 100 = 85.24% = 85% (2 sig. figs.) (1)	2
4 (d) (i)	the melting point will be the same as the literature/data book value (1) will have a sharp melting point/melt over a 1-2 °C range (1)	2
4 (d) (ii)	fingerprint region of sample and authentic compound are compared (1) exact match if compounds are the same (1)	2
4 (e) (i)	samples of the reaction mixture are removed and TLC run with sample of methyl benzoate (1) spot for methyl benzoate disappears when the reaction is complete (1)	2
4 (e) (ii)	singlet peak at 4.1 p.p.m. is for -O-C**H**$_3$ (1) singlet peak at 10.5 p.p.m. is for -COO**H** (1) neither peak is split because there are no Hs on an adjacent C atom (1)	3
4 (f) (i)	**Table 5** (see below) average titre = 29.80 cm^3 (1) moles of sulfuric acid = 0.0500 × 29.80/1000 (1) = 0.00149 Equation for the reaction: $2NaOH + H_2SO_4 \rightarrow Na_2SO_4 + 2H_2O$ (1) moles of NaOH remaining in 25.0 cm^3 = 2 × moles of sulfuric acid (1) = 0.00298 moles of NaOH remaining in 250.0 cm^3 = 10 × moles in 25 cm^3 (1) = 0.0298 mol original moles of NaOH = 1.60 × 25.0/1000 (1) = 0.0400 mol	

Table 5

Titration	1	2	3	4
Final/cm^3	32.7	30.15	31.05	31.40
Initial/cm^3	2.5	0.10	1.25	1.60
Titre	30.2	30.05	29.80	29.80

Question	Answer	Marks
	Equation for the reaction: $NaOH + C_6H_5COOH \rightarrow C_6H_5COONa + H_2O$ (1) moles of NaOH reacted with benzoic acid = 0.0400 – 0.0298 mol (1) = 0.0102 mol original amount of benzoic acid = 0.0102 mol (1) mass of benzoic acid in the crystals = moles × M_r = 0.0102 × 122 (1) = 1.2444 g purity = 1.2444/1.27 × 100 = 98.0 % (to 3 sig. figs.) (1)	11
4 (f) (ii)	error in use of the burette is ±0.05 cm^3 for each of two readings plus the error in judging the end-point to one drop ±0.05 cm^3, so total error is ±0.15 cm^3 uncertainty = 0.15/29.8 × 100 = 0.50 % (1)	1
4 (f) (iii)	uncertainty in balance = 0.01/1.27 × 100 = 0.8 % uncertainty in use of pipette = 0.12/25.0 × 100 = 0.5 % uncertainty in use of graduated flask = 0.6/250.0 × 100 = 0.24 % (1) the student was incorrect, greatest error is in the use of the balance (1) balance which measures to 3 d.p. could be used (1)	3
		Total 32
5 (a)	optical (1) an (asymmetric) carbon atom with four different groups attached (1) [a chiral molecular structure] COOH above C; HO, CH_3 and H attached to C (1) — HOOC above C; H_3C, OH and H attached to C (1) [the structures should show a mirror-image relationship] (equal but) opposite rotation (1) of plane-polarised light (1)	6
5 (b)	Isomer 1 and Isomer 2 are enantiomers (1) each is a non-superimposable mirror image of the other (1) a mixture of equal amounts of Isomer 1 and Isomer 2 is a racemate (1) optically inactive [50% (+) and 50% (–)] (1)	4
5 (c)	folding of protein gives an active site (1) which has a specific shape (1) which only fits one of the enantiomers as a key fits a lock (1)	3
5 (d)	*primary structure:* proteins are made up of aminoacids (1) the order/sequence: of these aminoacids is the primary structure (1) *secondary structure:* interactions between aminoacid residues in different parts of the chain (1) H-bonding between C=O••••H-N (1) *tertiary structure:* the folded shape of the protein (1) caused by interactions between residues, *e.g.* -S-S-, ionic attractions (1)	6

Question	Answer		Marks
5 (e)	**(i)** primary structure (1) **(ii)** β-pleated sheet (1) **(iii)** α-helix (1)		3
			Total 22
6 (a)	*meaning of hydrolysis:* breaking down with water *conditions:* (dilute) hydrochloric acid and heat	(1) (1)	2
6 (b) (i)	R_f of amino acids are 4.0/7.0 and 2.0/7.0 respectively = 0.57 and 0.29 so amino acids are methionine and glycine	(1) (1)	2
6 (b) (ii)	Different conditions cause the substances to move at different rates glycine value is closer than any other aminoacid R_f	(1) (1)	2
6 (b) (iii)	spray with ninhydrin uv light	(1) (1)	2
6 (c)	only hydrogen bonding is possible in (+)-lactic acid (+)-alanine exists as a zwitterion (dipolar ion) $H_3N^+CHCOO^-$ | CH_3 strong ionic bonding	(1) (1) (1) (1)	4
6 (d)	$H_3N^+CHCOOH$ to cathode | CH_3 $H_2NCHCOO^-$ to anode | CH_3	(1) (1)	2
6 (e) (i)	$H_2N-C(H)(CH_2SH)-C(=O)-N(H)-C(H)(CH_2SH)-C(=O)-O-H$ (1) for -CO-NH- link (1) for rest		2
6 (e) (ii)	*feature of cysteine:* thiol group/-S-H group *bond:* -S-S- *type of reaction:* oxidation	(1) (1) (1)	3

Question	Answer	Marks
6 (f)	$2RNH_2 + CH_3COCl \rightarrow RNHCOCH_3 + RNH_3Cl$ (1) (nucleophilic) addition-elimination (1) [Mechanism: $H_3C-C(=O)(Cl)$ + RNH_2 → $H_3C-C(O^-)(Cl)-N^+HR(H)$; labels (1), (1)] arrows and lone pair (1) structure (1)	6
		Total 25
7 (a)	*Equation:* $H_3C-CO-CH_2-CH_2-CH_3 + HCN \longrightarrow H_3C-C(OH)(CN)-CH_2-CH_2-CH_3$ (1) *Name of mechanism: nucleophilic addition* (1) *Mechanism:* [$H_3C-C(=O)-CH_2-CH_2-CH_3$ + $:C^-\equiv N$ → $H_3C-C(O^-)(CN)-CH_2-CH_2-CH_3$, H^+] (1) for each curly arrow starting at a bond or a lone pair (3) (1) for structure of anion *Name of product:* 2-hydroxy-2-methylpentanenitrile (1)	7
7 (b)	React with NaCN followed by dilute sulfuric acid. (1)	1
7 (c)	*Type of stereoisomerism:* optical (1) *Explanation:* carbonyl group is planar (1) attack by cyanide ion equally likely at either side (1)	3
		Total 11
8	**In this question answers will be marked on the level of response.**	
	Level 3: 5–6 marks All parts are covered and the explanation of each part is generally correct and virtually complete. Answer communicates the whole process coherently and shows a logical progression from part 1 and part 2 to overall explanation.	
	Level 2: 3–4 marks All parts are covered but the explanation of each part may be incomplete OR two parts are covered and the explanations are virtually complete. Answer is mainly coherent and shows a progression. Some statements may be out of order and incomplete.	

Question	Answer		Marks
	Level 1: 1–2 marks Two parts are covered but the explanation of each part may be incomplete and contain inaccuracies OR only one part is covered but the explanation is mainly correct and is virtually complete. Answer includes some isolated statements but there is no attempt to present them in a logical order or show confused reasoning.		
	Level 0: 0 marks		
	Part 1: ^1H NMR: butanone has 3 peaks, integration ratio: 3:3:2 butanal has four peaks, integration ratio: 3:2:2:1 *Part 2: infra-red:* butanone and butanal have same type of bonds so same absorptions above 1600 cm^{-1}, both have absorption between 1680 and 1750 cm^{-1} for C=O and absoption between 2750 and 3300 cm^{-1} for C–H *Part 3 precise molecular mass and 13C:* same molecular formula so same precise molecular mass butanone and butanal both have 4 peaks in 13C with similar chemical shifts		6
9 (a)	*Type of process:* hydrolysis or saponification *Reagent:* aqueous sodium hydroxide *IUPAC name of organic by-product:* propane-1,2,3-triol	(1) (1) (1)	3
9 (b) (i)	both R groups on same side of double bond *E*-configuration	(1) (1)	2
9 (b) (ii)	H_2 Ni catalyst	(1) (1)	2
9 (c)	transesterification $CH_3(CH_2)_{14}COOCH_3$ *or* $CH_3(CH_2)_{16}COOCH_3$ *or* $CH_3(CH_2)_7CH{=}CH(CH_2)_7COOCH_3$ drive equilibrium to right to oppose change imposed (Le Chatelier's principle)	(1) (1) (1) (1)	4
9 (d)	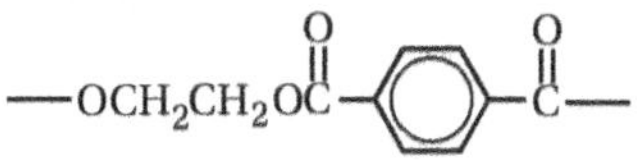 —$OCH_2CH_2OC(=O)$–C_6H_4–$C(=O)$— repeating unit can be split by hydrolysis by enzyme action	(2) (1) (1)	4
			Total 15
10 (a)	conc. H_2SO_4 + conc. HNO_3 $HNO_3 + 2H_2SO_4 \rightarrow NO_2^+ + H_3O^+ + 2HSO_4^-$ nitration electrophilic substitution 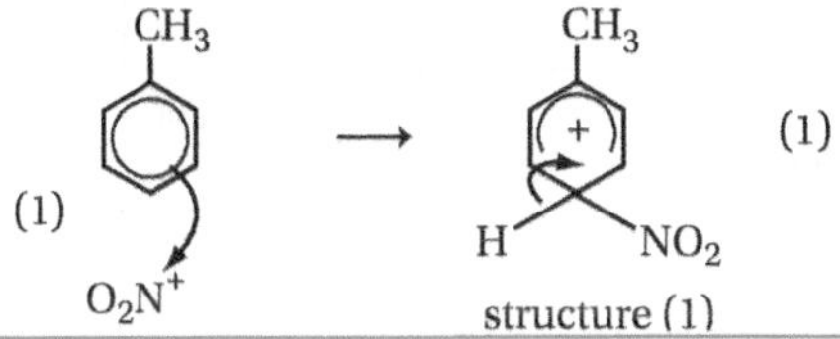 (1) methylbenzene (arrow from ring to O_2N^+) → intermediate with CH_3, ring (+), H and NO_2 (1) structure (1)	(1) (1) (1) (1)	7

Question	Answer		Marks
10 (b)	dissolve in a solvent, *e.g.* methylbenzene use column chromatography on alumina *or* silica collect main fraction remove solvent and recrystallise	(1) (1) (1) (1) (1)	5
10 (c)	oxidation	(1)	1
10 (d)	reduction *or* hydrogenation H_2 Ni catalyst	(1) (1) (1)	3
10 (e)	dissolve in aqueous HCl filter off compound **Q** add aqueous NaOH to precipitate **R** basic amino group in **R**	(1) (1) (1) (1)	4
10 (f)	esterification ethanol acid catalyst ethyl 4-aminobenzenecarboxylate	(1) (1) (1) (1)	4
			Total 24
11 (a)	$R\ddot{N}H_2 + H^+ \rightleftharpoons RNH_3^+$ inductive effect of ethyl group pushes electrons towards N increases electron density on N (lone pair more available) delocalisation of N lone pair in phenylamine reduces electron density on N (lone pair less available)	(1) (1) (1) (1) (1)	5
11 (b)	$[R_2N(CH_3)_2]^+Cl^-$ (1) (1) quaternary ammonium salt cationic surfactant	 (1) (1)	4
11 (c) (i)	*Explanation:* the neighbouring guanine base pairs bond to the platinum in cisplatin replacing the chloride ligands	(1) (1) (1)	3
11 (c) (ii)	-T-A-A-C-C-	(1)	1
11 (c) (iii)	hydrogen-bonding	(1)	1
			Total 14
12 (a) (i)	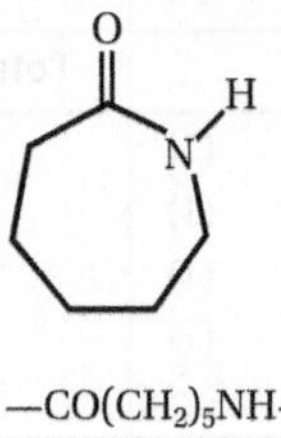 O H N $—CO(CH_2)_5NH—$	(1) (1)	2
12 (a) (ii)	$H_2NCH_2CH_2CH_2CH_2CH_2COONa$	(1)	1

Question	Answer		Marks
12 (b) (i)	*IUPAC name of **E**:* benzene-1,4-dicarboxylic acid *IUPAC name of **F**:* 1,4-diaminobenzene/benzene-1,4-diamine	(1) (1)	2
12 (b) (ii)	Polyamide	(1)	1
12 (b) (iii)	*Repeating unit of Kevlar* —NH—C_6H_4—NHCO—C_6H_4—CO— hydrogen bonding between chains forms a two-dimensional structure	(1) (1) (1)	3
12 (c)	paper *or* thin-layer chromatography comparison with known standards	(1) (1)	2
12 (d)	*Number of possible dipeptides:* 400	(1)	1
			Total 12
13 (a)	addition *or* chain-growth polymers —CH_2—CH(Cl)—	(1) (1)	2
13 (b)	carbon–carbon bonds are non-polar so inert cannot be split by biological organisms (enzymes)	(1) (1)	2
13 (c)	conserves natural resources (crude oil) recycled polymer may not be suitable for original purpose	(1) (1)	2
13 (d)	source of energy *or* reduces bulk liberates HCl	(1) (1)	2
			Total 8
14 (a)	gas–liquid *or* column chromatography	(1)	1
14 (b)	both have benzene rings, monosubstituted benzenes (or 1,4-disubstituted benzene)	(2)	2
14 (c)	C=O group uncoupled CH_3 group C=O group	(1) (1) (1)	3
14 (d)	$C_6H_5COCH_3$	(1)	1
14 (e)	alcohol OH group uncoupled CH_2 group	(1) (1)	2
14 (f)	$C_6H_5CH_2OH$	(1)	1
14 (g)	**G** cannot be oxidised by acidified potassium dichromate-no alcohol or aldehyde group **H** can be oxidised by acidified potassium dichromate - primary alcohol	(1) (1)	2
			Total 12
15 (a)	*step 1:* reagent : bromine conditions: uv light *step 2:* reagent : ammonia conditions: excess of ammonia	 (1) (1) (1) (1)	4

Question	Answer		Marks
15 (b)	*Name of mechanism:* electrophilic addition	(1)	
	(1) for each curly arrow (3) (1) for carbocation intermediate		5
15 (c)	*step 4:* reagent: potassium cyanide conditions: aqueous ethanol *Name of mechanism:* nucleophilic substitution	 (1) (1) (1)	
	(1) for each curly arrow (2)		5
15 (d)		(1)	
	reagent: H_2 conditions: Ni **or** reagent: $LiAlH_4$ conditions: ether as solvent	(1) (1) (1) (1)	3
15 (e)	overall yield route 1: 5% overall yield route 2: 61% *step 1:* low yield: bromine free radical is very reactive so substitutes every H atom it collides with *step 2:* further substitution occurs to give secondary and tertiary amines and quaternary ammonium salts	(1) (1) (1) (1) (1) (1)	6
			Total 23
16 (a)	*Reagent:* bromine (water) *Observation for cyclopentane* no visible change/stays orange *Observation for pent-2-ene* decolourised	(1) (1) (1)	3
16 (b)	*Reagent:* Tollens' reagent or Fehling's solution or acidified potassium dichromate(VI) *Observation for $CH_3CH_2COOCH_3$:* no visible change *Observation for $CH_3OCH_2CH_2CHO$:* silver mirror or red precipitate or goes green	(1) (1) (1)	3

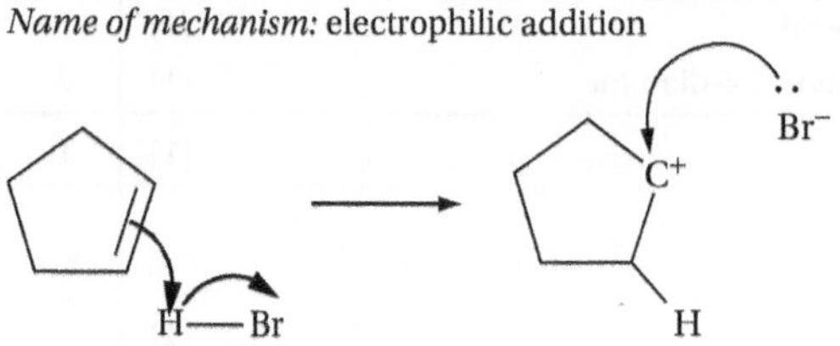

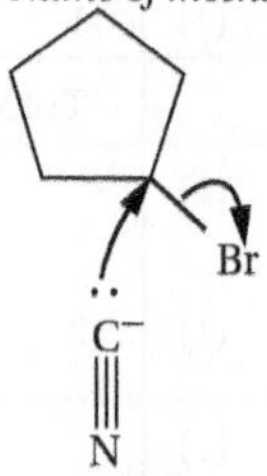

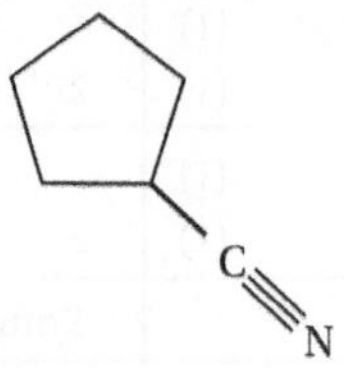

Question	Answer	Marks
16 (c)	*Reagent: e.g.* magnesium or sodium carbonate (1) *Observation for $CH_3COOCH_2CH_3$:* no visible change (1) *Observation for* $CH_3CH_2CH_2COOH$ effervescence (1)	3
16 (d)	*Reagent:* acidified potassium dichromate(VI) (1) *Observation for pentan-2-ol:* goes green (1) *Observation for 2-methylbutan-2-ol:* no visible change/stays orange (1)	3
16 (e)	*Reagent: e.g.* water (1) *Observation for propanoyl chloride* misty fumes (1) *Observation for chloropropane* no visible change (1)	3
		Total 15
17 (a)	 or *E*-isomer **L** 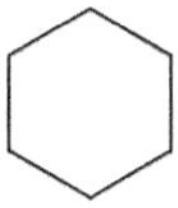 **M**	2
17 (b)	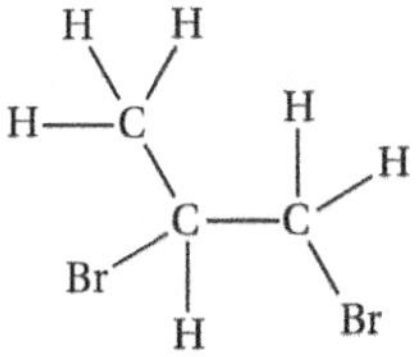 **P** 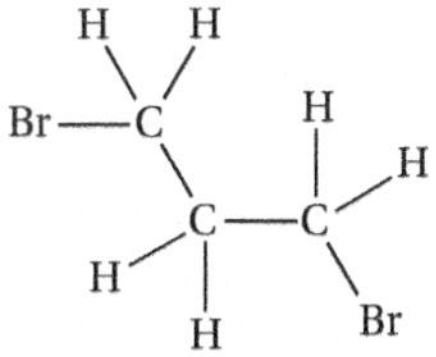 **Q**	2
17 (c)	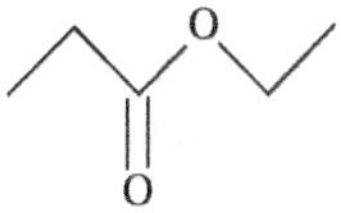 **R** 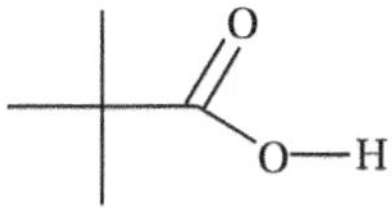 **S**	2
		Total 6

Multiple choice questions
1 A
2 D
3 D
4 C

Glossary

α-helix	a common type of secondary structure in a protein/polypeptide chain. The polypeptide chain forms a spiral for a specific length within the molecule. See also *β-pleated sheet*
β-pleated sheet	one of the two commonest types of secondary structure within a polypeptide or protein. See also *α-helix*
activation energy	the minimum energy required for a reaction to occur
active site	the location within an enzyme where a reaction takes place
acylation	the introduction of an acyl group into an organic molecule
acyl group	a functional group (RC = O) derived from a carboxylic acid
acylium cation	the electrophile $[RCO]^+$
addition polymer	one obtained by the addition of monomers to the end of a growing chain
adsorption	in chromatography, attachment of solute molecules to sites on the stationary phase
adsorption chromatography	involves a solid phase of finely-divided particles as the fixed (stationary) phase and a liquid or a gas as the moving (mobile) phase
amino acid	the name commonly used for compounds having a primary amino group attached to the carbon atom adjacent to a carboxylic acid group
analytical chromatography	operates with small amounts of material and aims to identify and measure the relative proportions of the various components present in a mixture
arenes	monocyclic or polycyclic aromatic hydrocarbons, such as benzene or naphthalene
aromatic	the name traditionally used in relation to benzene and its derivatives
Arrhenius equation	An empirical relationship, $k = Ae^{-Ea/RT}$, that expresses the manner in which the chemical reaction rate constant, k, varies as the reaction temperature changes
asymmetric carbon atom	one with four different atoms or groups attached that has no symmetry
atom economy	is a measure of how much of a desired product in a reaction is formed from the reactants
base (in DNA)	adenine, cytosine, guanine or thymine
base pair	bases linked by hydrogen bonding, e.g. adenine and thymine, between two adjacent strands of DNA
biodegradable	capable of being broken down by micro-organisms (enzymes)
biodiesel	a renewable, non-petroleum-based fuel obtained by transesterification from vegetable oils, such as rapeseed and soya bean oil
Brønsted-Lowry base	a proton acceptor
carrier gas	an *eluent* gas, such as helium, used as the moving phase in gas–liquid chromatography
catalyst	a substance that alters the rate of a reaction without itself being consumed
catalytic hydrogenation	the addition of H_2 to a multiple bond on the surface of a catalyst
chain-growth polymer	see *addition polymer*

chain isomers	structural isomers which occur when there is more than one way of arranging the carbon skeleton of a molecule
chain isomerism	occurs when there are two or more ways of arranging the carbon skeleton of a molecule
chemical shift (δ)	in NMR, the amount, measured in parts per million (ppm), by which a ^{1}H or a ^{13}C resonance, for example, is shifted from that of an internal standard
chiral centre	an atom that has four different groups bonded to it, see *asymmetric carbon atom*
chiral drugs	drugs possessing chiral centres, often single-enantiomer structures
chiral molecule	one that cannot be superimposed on its mirror image
chromatogram	a separated pattern of substances in a mixture, obtained by chromatography
chromatography	a technique for separating the components of a mixture on the basis of their different affinities for a stationary and for a moving phase
column chromatography	involves a stationery phase of finely-divided alumina or silica gel in a vertical glass tube and an organic solvent as the moving phase
complementary strand	either of the two chains that make up a double helix of DNA, with corresponding positions on the two chains being composed of a pair of complementary bases
condensation polymer	one involving the loss of small molecules, obtained by the reaction between molecules having two functional groups
delocalisation energy	the increase in stability associated with electron delocalisation
delocalised electrons	electrons that are spread over more than one atom in a molecule, e.g. as in benzene, where six delocalised electrons lie above and below the plane of the hexagonal ring
2-deoxyribose sugar	a five-carbon (pentose) sugar encountered as a component of DNA
deshielded	in NMR, a nucleus is said to be deshielded when the electron density surrounding it is reduced, giving rise to a downfield shift (larger δ value)
diazotisation	the conversion of $ArNH_2$ into NH_2^+
disulfide bridge	strong S—S bond formed in proteins by cross-linking between amino acid side chains; important in determining the tertiary structure of some proteins
DNA	deoxyribonucleic acid, a self replicating molecule that is a carrier of genetic information
DNA replication	the process by which a DNA double helix molecule is copied to produce two identical DNA
doublet	in NMR, a peak that is split into two parts
Ecoflex	a fully biodegradable aliphatic-aromatic co-polyester, used for disposable packaging, based on butane-1,4-diol and benzene-1,4-dicarboxylic acid
electrophile	an electron-seeking species, e.g. a positive ion or the more positive end of a polar molecule, which usually accepts a pair of electrons
electrophilic substitution reaction	mechanistically, an electrophilic addition-elimination reaction resulting in overall substitution, typically involving arenes, e.g. nitration of benzene
eluate	the solution emerging from a chromatographic column
eluent	the solvent used as the moving phase in column chromatography

elution	the process of washing the components of a mixture down a chromatographic column
enantiomers	three-dimensional, non-superimposable molecular structure mirror images
enzyme	a protein that acts as a catalyst for biological reactions
enzyme inhibitor	a molecule that reduces enzyme activity by binding to the active site
enzyme–substrate complex	the product of an interaction between an enzyme protein and a substrate molecule
***E–Z* stereoisomerism**	also known as geometrical or *cis–trans* isomerism
***E–Z* stereoisomers**	arise due to restricted rotation about a carbon–carbon double bond when the two pairs of attached substituents can be arranged in two different ways
fibrous proteins	contain long chains of polypeptides which occur in bundles, e.g. keratin
first order	the sum of the powers of the concentration terms in the rate equation = 1
fourth order	the sum of the powers of the concentration terms in the rate equation = 4
free-radical substitution reaction	one in which the hydrogen atom of a C—H bond is replaced by a halogen atom; a chain-reaction mechanism involves attack on a neutral molecule by a radical (halogen atom)
Friedel–Crafts acylation	an electrophilic substitution reaction, involving an acylium cation, resulting in carbon–carbon bond formation
functional group	an atom or group of atoms which, when present in different molecules, results in similar chemical properties
functional group isomers	structural isomers which contain different functional groups
functional group isomerism	occurs when different functional groups are present in compounds which have the same molecular formula
gas chromatography (GC)	involves an inert powder coated with a film of a non-volatile liquid, packed in a tube (the stationary phase), and a carrier gas (the moving phase)
glass-transition temperature (T_g)	the temperature at which a polymer changes from a hard and glass-like state to a more flexible and mouldable state
globular proteins	contain long chains of amino acids, soluble in water, which are folded into roughly spherical shapes, e.g. haemoglobin
in vivo	within the human body
incineration	waste-treatment technology involving the combustion of organic materials
initial rate of reaction	the rate of change of concentration at the start of a reaction
integration trace	in NMR, a computer-generated line, superimposed on the spectrum, which measures the relative areas under the various peaks in the spectrum
isoelectric point	the pH at which an amino acid has no net charge
isomers	molecules with the same molecular formula but in which the atoms are arranged differently (see *structural isomerism*)
Kevlar	a sheet-like polyamide, used in bullet-proof vests, derived from benzene-1,4-dicarboxylic acid and benzene-1,4-diamine
landfill site	an area of land on which rubbish is dumped
lock and key hypothesis	a mode of action which explains the specificity of enzyme catalysts

logarithm (natural)	a logarithm to the base *e*, where $e = 2.718$; $\ln(x) = \log_e(x)$
magnetic moment	a measure of the torque exerted on a magnetic system, e.g. a bar magnet, when placed in a magnetic field
mobile phase	see *moving phase*
moving phase	in chromatography, the liquid or gaseous phase that passes through a fixed stationary phase
multiplet	in NMR, a peak that is split into many parts
***n* + 1 rule**	in NMR, signals for protons adjacent to *n* equivalent neighbours are split into $n + 1$ peaks
nitration	the introduction of a nitro group (NO_2) into an organic molecule
nitryl cation	the electrophile $^+NO_2$
Nomex	the 1,3-linked isomer of Kevlar, used in flame-resistant clothing
nuclear spin	a property that influences the behaviour of certain nuclei, typically ^1H and ^{13}C, in a magnetic field; nuclei possessing even numbers of both protons and neutrons, such as ^{12}C and ^{16}O, lack magnetic properties and do not give rise to NMR, signals
nucleophile	an electron pair donor; a species that is strongly attracted to a positive centre
nucleophilic addition reaction	one in which an electron-rich molecule or ion (with a lone pair of electrons) attacks the electron-deficient atom of a polar group, e.g. overall addition of HCN to an aldehyde or ketone
nucleotide	a subunit of nucleic acids like DNA (deoxyribonucleic acid), composed of a nitrogenous base, e.g. adenine, a five-carbon (pentose) sugar, e.g. 2-deoxyribose, and at least one phosphate group
optical isomers	stereoisomers (enantiomers) which rotate the plane of plane-polarised light equally but in opposite directions
optical isomerism	compounds whose molecules, although alike in every other way, have structures that are mirror images of each other due to an asymmetric carbon atom. This chiral centre causes a difference in the molecules' effect on plane polarised light (single colour) light
optically active	capable of rotating the plane of plane-polarised light
order of reaction	the sum of the powers of the concentration terms in the rate equation
paper chromatography	involves a thin layer of water adsorbed on to chromatographic paper (the stationary phase) and a solvent or solvent mixture (the moving phase)
partition	in chromatography, the equilibration of solute molecules between stationary and moving phases
partition chromatography	involves a thin, non-volatile liquid film held on the surface of an inert solid or within the fibres of a supporting matrix (the stationary phase) and a liquid or a gas (the moving phase)
parts per million	just as per cent means out of a hundred, so parts per million (ppm) means out of a million
peptide link	the amide bond (CO-NH) in polypeptides and proteins
plasticiser	a substance used to soften plastics and increase flexibility, e.g. dibutyl benzene-1,2-dicarboxylate

position isomers	structural isomers which have the same carbon skeleton and the same functional group(s), but in which the functional groups are joined at different places on the carbon skeleton
position isomerism	occurs when the isomers have the same carbon skeleton, but the functional group is attached at different places on the chain
preparative chromatography	a form of purification of organic compounds, involving chromatography on a relatively large scale
primary structure	of a protein, is the sequence of amino-acid units present in the polymer
protein	a large molecule made from one or more polymer chains of amino acids linked by peptide bonds
proton-decoupled spectra	are simplified NMR spectra obtained as the result of removing the interactions between ^{13}C nuclei and any attached protons
quartet	in NMR, a peak that is split into four parts
racemate	a mixture containing equal amounts of both enantiomers
racemic mixture	see *racemate*
rate constant (*k*)	the constant of proportionality in the rate equation
rate-determining step	the slowest step in a multi-step reaction sequence
rate equation	the relationship between the rate of reaction and concentrations of reactants
rate of reaction	the change in concentration of a substance in unit time
reaction mechanism	a sequence of discrete chemical reaction steps that can be deduced from the experimentally observed rate equation
recycling	the processing of used materials, e.g. glass, paper, textiles, metals and plastics, into new products in order to prevent wastage, to reduce the consumption of raw materials and to lower energy costs
relative intensities	in NMR, the areas under the various peaks in the spectrum
repeating unit	a part of a polymer which periodically repeats itself along the polymeric chain
replication	the process of copying DNA molecules during cell division. The two strands unwind and act as templates for the creation of two identical copies
resolution	the separation of enantiomers *or* the separation of mixtures of chemicals using chromatography
resolved	in paper chromatography, complete separation of the components of a mixture in one direction
resonance	a concept used when a single molecule can be approximated by more than one structure involving single and multiple covalent bonds *or* in NMR, the excitation of atomic nuclei in a magnetic field by exposure to electromagnetic radiation of a specific frequency
resonance energy	the increase in stability associated with resonance between Lewis structures
resonance hybrid	a representation of an actual molecule, e.g. benzene, when classical structures using single and multiple covalent bonds are inadequate
retention factor (R_f)	in chromatography, $R_f = \dfrac{\text{distance travelled by the compound}}{\text{distance travelled by the solvent front}}$

retention time	in chromatography, the time each component remains in the column
second order	the sum of the powers of the concentration terms in the rate equation = 2
secondary structure	of a protein, relates to the orderly, hydrogen-bonded arrangements between peptide chains resulting in either a helix or a pleated sheet
shielded	in NMR, a nucleus is said to be shielded when the electron density surrounding it is increased, giving rise to an upfield shift (smaller δ value)
singlet	in NMR, a peak that is not split
solvent front	in paper chromatography, the position reached by the leading edge of the solvent after separation has occurred
spin–spin coupling	in NMR, the interaction between the nuclear spins of non-equivalent hydrogen atoms on adjacent carbon atoms
splitting	in NMR, the splitting of an absorption signal (a peak) into more complex patterns as a result of coupling between neighbouring nuclear spins
stationary phase	in chromatography, the fixed phase through which passes the moving or mobile phase
step-growth polymer	see *condensation polymer*
stereoisomerism	occurs when molecules with the same structural formula have bonds arranged differently in space (see *E–Z stereoisomerism* and *optical isomerism*)
stereoisomers	are compounds which have the same structural formula but have bonds arranged differently in space
stereospecific	the requirement that only one stereoisomer of a chiral substrate can bind successfully to an active site on an enzyme
structural isomerism	occurs when the component atoms are arranged differently in molecules having the same molecular formula
structural isomers	compounds with the same molecular formula but different structures
substrate	the reactant in an enzyme-catalysed reaction
sugar	a simple carbohydrate; for example, glucose, sucrose, maltose
surfactant	a wetting agent, containing hydrophobic and hydrophilic groups, able to lower the surface tension of a liquid and the interfacial tension between two liquids; the name is derived from surface acting agent
tertiary structure	of a protein, relates to the overall three-dimensional shape of the protein
Terylene	a polyester, used in permanent-press fabrics, derived from benzene-1,4-dicarboxylic (*terephthalic*) acid and ethane-1,2-diol
thin-layer chromatography (TLC)	involves a thin layer of a polar, adsorbent material coated on to a glass plate or on to an aluminium or plastic sheet (the stationary phase) and a solvent (the moving phase)
third order	the sum of the powers of the concentration terms in the rate equation = 3
transesterification	a reversible reaction in which an ester reacts with an alcohol, usually in excess, to form a new ester and a new alcohol
triplet	in NMR, a peak that is split into three parts
zero order	the sum of the powers of the concentration terms in the rate equation = 0
zwitterion	a dipolar ion that has both a positive and a negative charge, especially an amino acid in neutral solution

Index

Notes

Notes

Notes

Notes